Arthur Joseph Clay

A Manual of Linear Shorthand

An original scientific alternating system

Arthur Joseph Clay

A Manual of Linear Shorthand
An original scientific alternating system

ISBN/EAN: 9783337418847

Printed in Europe, USA, Canada, Australia, Japan

Cover: Foto ©berggeist007 / pixelio.de

More available books at **www.hansebooks.com**

A

MANUAL

OF

LINEAR SHORTHAND.

– ··⁑·· –

AN ORIGINAL SCIENTIFIC ALTERNATING SYSTEM.

— ·⁑·· –

BY

A. J. CLAY, M.A., Oxon.

—··⁑·—

PART I.—CORRESPONDING STYLE.

—··⁑·—

LONDON:
BEMROSE & SONS, LIMITED, 23, OLD BAILEY;
AND DERBY.
—
1898.

PREFACE.

THIS system is described as an original Scientific Alternating System. Perhaps these terms require some explanation.

Linear Shorthand was not founded, as most of its predecessors have been, upon some older system. It is based directly upon investigations undertaken by the author for the purpose of certain papers read before the Shorthand Society of London, under the title "The Science of Shorthand." These papers, which form the introduction to the present volume, were three in number, and dealt with an Analysis of the Sounds of Speech, an Analysis of the Signs of Writing, and with the rules and principles of their combination. Very early in the preparation of the second of these papers, in 1894, the author was led to the conviction that the best basis for a system of Shorthand was the "alternating" method, and he was led to abandon a system of the "cursive" or "script-geometrical" type which he was then using, and to formulate the alphabet and leading rules of Linear Shorthand. These were found to bear out the correctness of the principles upon which they were based by the manner in which they worked out in practice, and during the three years that the system has been in constant use, no change has been found necessary except in the smallest details. The Alternating method depends, as is explained on page 13, upon the alternation of significant downstrokes, or characters, and non-significant upstrokes, or liaisons. In Longhand, and in the English systems of Bordley, Roe, Oxley, Adams, &c., the upstrokes have no meaning whatever, and therefore the writing is somewhat lengthy. Gabelsberger in his system, published in Germany in 1831, first showed how the upstrokes, though unsuited to bearing any meaning in themselves, might be used to indicate the vowel sounds by modifying the relative positions of the consonants. The existing systems, however, formed on this plan are so irregular and unscientific that the principles which really underlie the method are rather concealed by them than elucidated, and it was therefore necessary to proceed *ab initio*, and discover, if possible, what these principles were, before the foundations of Linear Shorthand could be laid down.

The true principles then of an Alternating Shorthand System seem to the Author to be as follows :—

(i.) That it is possible to construct a practical Shorthand system in which :—

(a) All the consonants shall be represented by simple downstrokes.

(b) All the vowels shall be indicated by the help of the connecting upstrokes.

(c) All consonant combinations shall be shown without the intrusion of an upstroke.

(ii.) That such a system, as compared with one constructed upon the usual " articulate " lines, will be :—

(a) More facile, because the writing will be, like Longhand, on one slope, and will not admit awkward junctions.

(b) Briefer, because the number of concurrent differentiations which can be employed is greater.

(c) More legible, because fuller vocalization is possible.

It would take too much space to set forth fully how the German authors have departed from these principles, but, for those who are students of Shorthand Science, a list is appended of the chief points where the systems of Gabelsberger and Stoltz, as adapted to English by Richter and Dettmann, diverge from the scientific type :—

(i.) Consonants represented by upstrokes :—

Gabelsberger - t, f, p, v.

Stoltz - - - t, th.

(ii.) Consonants represented by horizontal strokes, that is, signs whose termination is on the same level as their commencement :—

Gabelsberger - l, th, s, n.

Stoltz - - - r, l.

(iii.) Vowels represented by significant characters :—

Gabelsberger - u, ür, i, o, aw.

Stoltz - - - (In the latest edition, none).

It is easy to trace through the systems the difficulties caused by these departures from type, and the special rules which they necessitate. There are also irregularities, more numerous in Gabelsberger than in Stoltz, in the tabulation of the alphabet, in the vowel rules, and in the construction of Consonant groups, but space does not allow of their being enlarged upon. The author hopes that what has been said is sufficient to show that the principles of alternating shorthand systems have never been recognised or followed, that Linear Shorthand was founded, not upon a previous system, but upon these hitherto unformulated principles, and that the formulation of them which is given above, and the construction thereupon of a practical system, constitute an original and important addition to Shorthand Science.

In learning this system the student is advised to pay more attention to reading than to writing. The reverse is usually recommended, probably owing to the expense of shorthand specimens. It is hoped that the number of examples given in this book may lead to complete uniformity among all the writers of the system. The course recommended to the learner is as follows:—First, read carefully through the Introduction, which explains many things which might otherwise cause some difficulty. Then proceed to the Lessons, and master each one thoroughly before turning to the next. Very great pains have been taken in the construction of the Exercises, which will be found to form a special feature of the book. They must be conscientiously performed; a careful attention and steady diligence during the duller parts of the commencement, will lead to a certain and rapid success when the more interesting portions of the book are reached. Before commencing a new Lesson, read over the Shorthand in the three preceding exercises; then read carefully the new lesson, copying on paper and pronouncing aloud all the examples; then work through the new exercises as directed. By devoting to the study of Shorthand two hours a day, the learner should before the end of a month be able to write correctly and neatly any passage at about thirty words a minute (rather faster than most persons can write legibly in Longhand), and to read his Shorthand without any difficulty. He must then begin to practice speed. The only method is to write from dictation. Books may be obtained from the Publishers containing speeches marked to be read at various speeds. Thirty hours' practice will enable most persons to increase their speed to sixty words a minute, and thirty more, to eighty words. Then, but not before, the learner should purchase the second part of the Linear Shorthand Manual, which contains the Reporting Style. With the aid of this, and continued practice, he will be able to attain to any speed desired. It is to be specially noticed that there is no difficulty in reading Linear Shorthand, beyond that of the novel characters; words are represented just as they are sounded, and not represented by the bare consonant skeletons which make the deciphering of some systems so difficult.

In order to afford reading practice to learners of Linear Shorthand, the
2

"Pickwick Papers" are being published in monthly parts: each number consists of sixteen pages, and contains facsimile shorthand and transcript in parallel columns. The price has been fixed at sixpence a number, and it is hoped that all learners of the system will subscribe to this publication to enable it to be continued.

In conclusion, the Linear Shorthand Company *earnestly request* every purchaser of this book to send them the enclosed postcard, stating his name and address, when he will be kept informed, from time to time, of the progress of the system.

INTRODUCTION.

THE SCIENCE OF SHORTHAND.

Being a Reprint of Papers read by the Author before the Shorthand Society of London.

PART I.

ANALYSIS OF THE SOUNDS OF SPEECH.

THE Science of Shorthand may be divided, like other Sciences, into two main branches—Analysis and Combination. The first step is to analyse the materials with which the Science deals, and the second to investigate the laws of their combinations. The materials with which the Science of Shorthand deals are the Sounds of Speech and the Signs of Writing. The present paper is an attempt to analyse the first of these.

An analysis of the Sounds of Speech of all languages would be quite unnecessary for the purpose of English Shorthand, and would, moreover, be beyond my powers. I do not intend to do more than analyse exhaustively the sounds of English Speech, but shall be compelled to notice a few foreign sounds to illustrate my remarks.

Speech is the communication of Ideas, by differentiation of Voice.

Voice is the peculiar sound caused by the vibration of the Vocal Chords; it is differentiated by movements in the Oral Cavity.

To understand the sounds of English Speech it will be necessary to examine first the production of voice, and secondly the modifications of it in the mouth.

Voice is ordinarily produced by an *exhalation* of breath, though it may also be produced by an *inhalation*. The air passes from the lungs by the Trachæa, or Wind Pipe (*a*, Fig. I), and between the Vocal Chords (*b*), which are stretched across a species of chamber called the Larynx (*c*). By bringing the Vocal Chords nearly into contact they may be made to vibrate, which gives rise to what is called Voice. The air next passes in a backward direction into the upper part of the Oesophagus, or Gullet (*d*), where it widens into the Pharynx (*e*). The opening through which it passes may be closed when necessary by the Epiglottis (*f*), which curves downwards and cuts off the communication between the Oesophagus and Trachæa. From the Pharynx the breath issues either through the Oral Cavity (*g*), or through the Nazal Cavity (*h*). It is not necessary to go into the question of *pitch*, beyond saying that this seems to be determined, (i.) by the tension of the Vocal Chords; (ii.) by the shape of the Trachæa; (iii.) by the position of the Epiglottis and Uvula (*k*); and (iv.) by the size and shape of the Pharynx. Pitch, besides giving musical power to the voice, is the means whereby we introduce *expression* in speaking, and are enabled so exactly to indicate the emotions.

Fig. I.

  a. Trachæa ("Wind Pipe").
  b. Vocal Chords.
  c. Larynx.
  d. Oesophagus (Gullet).
  e. Pharynx.
  f. Epiglottis.
  g. Oral Cavity.
  h. Nazal Cavity.
  k. Uvula.
  m. Hard Palate.
  n. Hanging Palate.

Lower Organs.		Upper Organs.	
1. Lip.		6. Lip.	
2. Point		7. Teeth.	
3. Front	of	8. Back of Teeth.	
4. Centre	Tongue.	9. Point	
5. Back		10. Front	of
		11. Centre	Palate.
		12. Back	

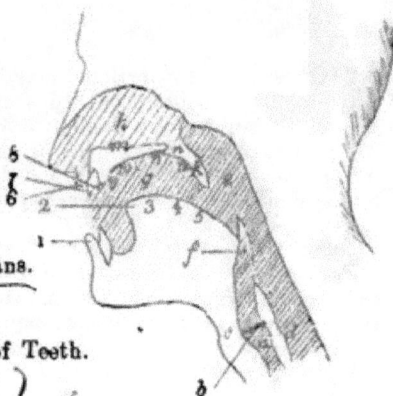

It is in passing through the Oral Cavity that articulate character is given to voice. The changes here produced in it may be divided into two kinds, Vowel Modification, which is produced by mere alteration of the size and shape of the mouth, and Consonant Modification, which is produced by contact or semi-contact between different parts of its floor and roof. The parts which come into contact, or nearly so, to form the various consonants, may be considered to be twelve in number, and for the sake of convenience I venture to call them the "Organs" of Speech. The lower organs, which are movable, are five in number (see fig. I); the Lip (1); the Point of the Tongue (2); the Front, or the part between the point and the centre (3); the Centre (4); and the Back of the Tongue (5). The upper organs, which are fixed, are seven in number; the Lip (6); the Teeth (7); the Back of the Teeth (8); the Point of the Palate, or the projecting ridge caused by the gums (9); the Front, or part between the point and the centre (10); the

Centre (11) ; and the Back of the Palate (12). Most of the lower organs move up into contact or semi-contact with the upper organ which is opposite to them, but two of them, the Lip and Point of Tongue, move back and forward into contact with others of the upper organs, beside their immediate opposites.

Having thus briefly sketched the manner in which voice is produced and modified, I proceed to the first of the two kinds of modification noticed, namely, Vowel modification. In sounding the vowels the emission of the voice is almost free or uninterrupted, but its character is changed in some rather obscure manner by the alteration in shape and size of the Oral Cavity. The number of possible vowels is theoretically infinite. It would, therefore, be useless to attempt an analysis of the means by which they are differentiated ; all that can be done is to examine and classify those used in correct speech.

The main difference between vowels is that between Simple and Compound. Simple Vowels are those during the pronunciation of which the shape and size of the mouth-cavity remain unaltered. Compound Vowels are formed by commencing the emission of voice with the mouth in the same position as for one of the Simple Vowels, and causing it *gradually* to assume one of three positions which are nearly, but not quite, identical with those for the Simple Vowels, I, ŭ, ōō.

There is a second, but much less important difference, that between short and long. As generally applied to vowels these words are not used in their ordinary sense, from which unfortunate circumstance a great deal of confusion has arisen. I use the terms in their strict and natural sense, indicating the time during which the utterance of the vowel is prolonged. Some vowels are used in three lengths, as **aw** in *salt, sought,* and *sword* ; some only in two, as **ay** in *pate* and *paid* : some only in one, as **é** in *met*.

The simple vowels which we will consider are nine in number.

ă is found only in the short form, as in *but, can,* &c.

ĕ is found in English only in the short form, as in *bet, Nell,* &c. ; but it is commonly prolonged in other languages, as in the French, *bête, guère,* &c.

ĭ is found in English only in the short form, as in *bit, filled,* &c.

ŏ is only found in the short form, as in *rot, moss,* &c.

ŭ is found in the short form in words like *nut, shun,* &c. The long form, **ur**, is not quite a simple prolongation of **ŭ**, though it approaches very near to it.

oo is found in the short form in *put, full,* &c. It is not used in the long form in English, though the compound **oo** approaches very near to being a simple sound.

aw occurs in its shortest form only before **l**, followed by a surd consonant, as in *malt, false,* &c. A longer form is found in *sought, daughter,* &c. ; and a still longer form in *laud, cause,* &c. Most writers have failed to notice the short form in *salt, false,* &c., and have said that the long **aw** in *sought, torse,* &c., is a prolongation of the **o** in *sot, toss*.

o occurs in a short form in the words *only, home, wholesome,* &c. It is not found in a long form in English, but **is** common in Continental languages in both forms, as in *homme, col, mot* (short) ; *rôder, clôture, hôte* (long).

a does not occur in English. It is common in French, as in *la, patte* (short) ; and *râle, garde, débâcle* (long). It is noticed here because it forms the base of two English Compound Vowels.

We may now pass to the consideration of the Compound Vowels. As above stated, during the pronunciation of the Compound Vowels, the size and shape of the mouth cavity is slowly altered from that necessary for the production of one of the simple vowels to that of one of three terminations, which approximate very nearly to, but are not identical with the three simple vowels I, ŭ. oo.

It is convenient to speak of a compound as formed on the *base* of a simple vowel by the addition of a *termination*; but it must be understood that the compound vowels are not mixtures, but true compounds; that is, they are formed not by adding one sound to another, but by a continuous transition from the oral position of the first to the oral position of the second.

Fig. II.

Simple Vowels forming Bases.	Terminations.		
	ĭ	ŭ	ŏŏ
Vowel—as in	Compound—as in	Compound—as in	Compound—as in
ă pat		aîr pair	
ĕ pet	ay pate	eer ne'er	
ĭ pit	ē peat	eer peer	
ŏ pot		ah part	
ŭ but		ur pert	
ŏŏ put		oor poor	ōō boot
aw fault	oy boy	ore four	
ô wholesome		o'er Noah	ô know
[å] la, patte (Fr.)	i bite	ire fire	ow pout

The formation of the Compound Vowels will be found in Fig. II., but I should like to say a few words about one or two of them.

Most of the compounds closing with the ŭ termination are represented in the ordinary spelling by combinations containing the letter r, as *air, near; fur, fur.* In English pronunciation the r is not as a rule sounded, and should not therefore be written.

The vowel **ah** is generally left unprovided with a corresponding short vowel, or is said to be a prolongation of **ăt.** It is not far removed from a simple prolongation of **ŏt,** compare **cot: cart,** but there is a slight difference, which shows that it is a compound and not a simple vowel.

The three vowels, i, as in *bite,* ire, as in *fire,* and ow, as in *pout,* are shown in Fig. II. as compounds with the base â, as in *la, patte* (Fr.), and the terminations I, ŭ, and oo respectively. This view is so different from that generally taken that I do not expect it to be received at first, but remain assured of its correctness.

It is necessary to distinguish those compounds formed by the bases, or simple vowels, with the termination ŭ, from the double vowels or diphthongs (properly so called), formed by the addition of the same termination to the compound vowels terminating in i and oo. Such diphthongs are found in words like :—*slayer, mower, knower, buyer,* and are quite distinct from the compound vowels in :— *we'er, peow, Noah, fire.*

Most of the compound vowels are used in both short and long forms, as in *boat-load*; but since they take as a rule rather longer to pronounce than simple vowels, we are accustomed to call most of the former long, and most of the latter short.

The above is all that need be said here about the compound vowels. Those to which I have not referred will be found in Fig. II., which, I think, almost explains itself. Before passing, however, to the consideration of the consonants, it is necessary to mention the *neutral* vowel, which occurs in such words as *battle, taken, parrot, retour, contretemps.* This vowel sound is caused by the passing of the organs of the mouth, by the easiest transition, from the position necessary for the formation of the preceding, to that necessary for the succeeding consonant. In careless English the majority of the vowels are so slurred as to be nearly indistinguishable from the neutral vowel, a fact which is of great importance to the shorthand inventor. It renders possible the system explained in Lesson XIV., and there called the "neutralization" of the vowels.

I pass now to the second of the two kinds of voice-modification, namely, Consonant modification. This is produced by contact or semi-contact between the various parts of the roof and floor of the mouth, to which I have applied the name of Organs. The differences between Consonants are of three kinds : Organic, Mechanical, and Vocal. They are shown in Fig. III.

Fig. III.

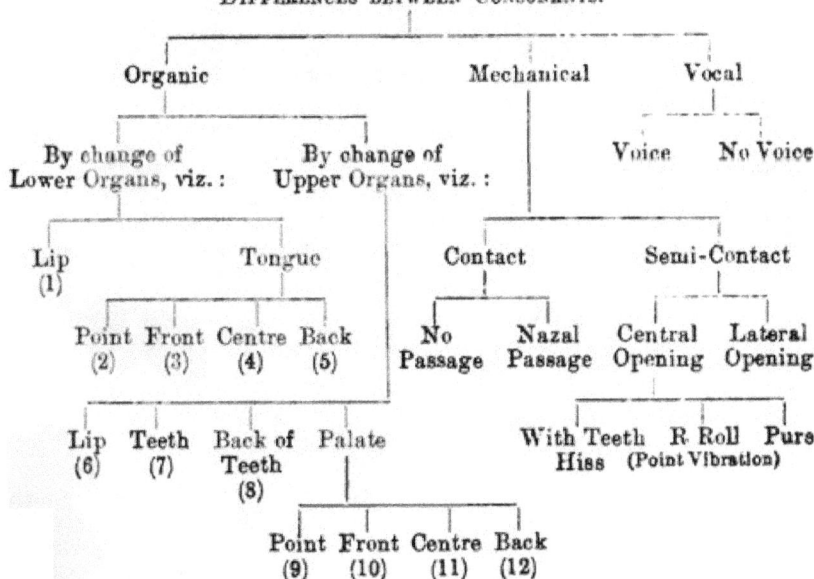

DIFFERENCES BETWEEN CONSONANTS.

Organic differences are produced by a change of the forming Organ. It is obvious that it is impossible for any of the lower, or mobile organs to touch *all* the upper or fixed organs. In fact, as has been said, only two of the lower organs come into contact with other upper organs than their immediate opposite, and the organic differences in English are, as will be seen in the first column of Fig. IV., eight in number. Instances of organic differences are those between b and d. s and t, f and th.

Mechanical differences are produced by difference of the mechanical nature of the contact between the forming organs. This may be perfect or partial, contact or semi-contact. In the case of perfect contact, the emission of air from the mouth may either be momentarily stopped altogether, and then allowed to escape with, as it were. an explosive effect, or it may be given a passage through the nazal cavity. The perfect contact is observed in the consonants p, t, and k: the nazal passage in m. n, and ng. In the case of semi-contact, where the organs are *partially* in contact, the air may escape either through a medial or through a lateral opening between them. In the former case the articulation may be pure, as in the case of y, or it may be accompanied by the teeth-hiss, as in s, th, f, and sh, or by the point-vibration, as in r. Of the lateral opening, the only instance in English is l. The six mechanical differences combine with the eight organic differences to form fifteen varieties, which are shown in Fig. IV. Instances of mechanical differences are those between p and m; l, z, n, and d; k and kh.

Vocal difference is produced by the presence or absence of voice (that is, of vibration of the vocal chords) at the moment when the contact or semi-contact takes place. Instances of consonants produced without vibration of the vocal chords are p, t, and k; instances of consonants produced with vibration of the vocal chords are b, d, and g. The former are called Surd, the latter Sonant Consonants. In the pronunciation of Surd Consonants, at the instant of the contact or semi-contact, the vocal chords cease to vibrate, the vibration commencing again as soon as the organs separate for the formation of the ensuing vowel, while in the pronunciation of Sonant Consonants the vocal chords are continuously vibrating. The above explanation is generally considered to be complete, but if it were so, it is evident that in *whispering*, when there is no vibration of the vocal chords at all, it would be impossible to distinguish between Surd Consonants and Sonant. This is not the case, and it is evident therefore that there is some other difference. This is probably to be found in a more forcible contact between the forming organs in the case of surds, so that when the contact or semi-contact is broken, the pent up air escapes with more violence, with a greater explosion, than after the corresponding sonant consonants. There is another fact to be observed in connection with vocal difference, namely, that vowels are longer before sonants than before surds (the term long is still used in its literal meaning, as explained above). The reason for this is obvious. As has been shown, the vibration of the vocal chords is stopped at the moment of contact or semi-contact in forming a surd consonant. In an unconscious anxiety to avoid carrying on the vibration over the instant of organic contact, we are accustomed to cut the vowel a little shorter before a surd than before a sonant, as will be seen from observing the vowel in *pate* and *paid*.

If each of the fifteen Organo-mechanical varieties was used in both the surd and sonant forms, we should have thirty distinct consonant sounds in

English. I do not, however, consider that the nasals m, n, and ng, the so-called liquids r and l, and the semi-vowel y, can be used in the surd form. Some writers think that the first five of these, at any rate, are so used, when they "coalesce" with another consonant which is a surd, as in *camp, sent, rank, pray, fault.* With this view I venture to disagree, considering that the vibration of the chords does not in correct speech cease in such words, till the organs assume the position of the surd consonant, with which the m, n, l, &c., are combined. The number of distinct consonant sounds used in English is therefore twenty-four; they are shown, with the means of their differentiation, in Fig. IV.

Fig. IV.

Organic Difference.		Mechanical Difference.		Vocal Difference.	Consonants as in
Lower	Upper			Surd	P	Pope
Organ		Contact	No Passage	Sonant	B	Babe
	Lip		Nazal Passage	Sonant	M	Mumps
Lip		Semi Contact Pure		Surd	Wh	Which
				Sonant	W	Witch
	Teeth	Semi Contact Teeth Hiss		Surd	F	Fife
				Sonant	V	Vivid
Point of Tongue	Teeth	Semi Contact Teeth Hiss		Surd	Th	Three
				Sonant	Dh	Thee
	Back of Teeth	Contact	No Passage	Surd	T	Fight
				Sonant	D	Died
			Nazal Passage	Sonant	N	Nance
		Semi Contact	Teeth Hiss	Surd	S	Sauce
				Sonant	Z	Ways
			Lateral Opening	Sonant	L	Lilt
	Point of Palate	Semi Cont. Point Vibration		Sonant	R	Pry, Ripe
Front of Tongue	Front of Palate	Semi Contact Teeth Hiss		Surd	Sh	Vicious
				Sonant	Zh	Azure
Centre Tongue	Centre Palate	Semi Contact Pure		Sonant	Y	You
Back of Tongue	Back of Palate	Contact	No Passage	Surd	K	Cork
				Sonant	G	Gag
			Nazal Passage	Sonant	Ng	Song
		Semi Contact Pure		Surd	Kh	Loch
				Sonant	Gh	Lough

In the ordinary alphabet there are only twenty-one consonant forms, and these only represent sixteen consonant sounds. The five letters which do not stand for distinct consonant sounds are C, H, J, Q, and X. Of these H expresses the aspirate, which is not a consonant, and which we have yet to consider; C is used only as an alternative sign; J, Q, and X stand for combinations. Since then there are only sixteen distinct consonant signs in the ordinary alphabet, it is obvious that there are eight consonant sounds, which are not provided with signs. I proceed to detail how these are represented. They merit our attention because we have been so long accustomed to represent them improperly that we have probably formed erroneous ideas about them.

(i.) The surd sound of **w** is generally written Wh ; phonetic reformers have written it Hw. Neither is really correct ; the sound when correctly given differs from **w** as f differs from **v**. During the pronunciation of surd **w** and of **f**, the vocal chords are not vibrating, during the pronunciation of **w** and **v** they are vibrating.

(ii.) and (iii.) The sounds which begin the words *the* and *three* are both represented by the combination Th, which is also used in its correct sense in words like *pothook*. They have the same relation to **t** and **d**, as f and **v** have to **p** and **b**, and require special characters as much as these.

(iv.) and (v.) The sound generally represented by Sh is found in many words, and is sometimes represented by ti, si, and sci, as in *rush*, *ration*, *Asia*, *conscience*. The sound represented by J in French occurs fairly often in English, and is generally represented by s, si, or z, as in *treasure*, *vision*, *azure*.

(vi.) The sound generally represented by Ng is a very common one in English, and is sometimes indicated by n only before another gutteral, as in *bank*, *anger*.

(vii.) and (viii.) The gutteral sounds heard in the Scotch *loch*, and Irish *lough*, are not used in modern English, but are necessarily introduced into a scheme of English consonants. They bear the same relation to wh and **w**, that **k** and **g** do to **p** and **b**.

We have now examined as fully as space permits the vowels and consonants used in English. There is one other sound, which is neither a consonant nor a vowel. This is the aspirate **h**, a pure emission of unvocalized breath, it is in fact a *surd vowel*. It is always followed by a vowel, and the shape necessary for the vowel is assumed before the emission of the aspirate. "Dropping an h" means that the vowel is vocalised too soon. It is a curious fact, and one worth repeating here, that in the earliest times aspirates were dropped by the uneducated ; in fact, medial aspirates, as in our *behind*, and the old Greek ταὼς (*peacock*), had probably disappeared from the Greek language before classical times.

There are, therefore, forty-nine sounds distinguished in English, eight simple vowels, sixteen compound vowels, twenty-four consonants, and the aspirate. The force of habit is so strong, and in the domain of phonetics it has been exerted so constantly in the direction of error, that it is very difficult, without long practice, to discover what we do actually speak ; and even after long practice mistakes are still so easy than an absolutely correct analysis of the sounds of speech will never be made. It is hoped that the above may be considered in some measure an advance upon its predecessors, and that if any of the statements contained in it appear at first sight novel and unnatural, they may not be taken as being therefore false, but may be carefully and thoughtfully considered before they are rejected.

PART II.

ANALYSIS OF THE SIGNS OF WRITING.

This is by far the most difficult portion of my subject, and that for two reasons, first, because the number of Signs that may be made seems to be almost innumerable, and secondly, because it is very difficult to prevent oneself going into the question of Signification, which is quite foreign to the analysis. Owing to these difficulties such an analysis has never been carried out with even an approach to completeness, and but rarely attempted. The result is that systems are either founded upon an incomplete analysis on geometrical lines, or else they are drawn up without any scientific basis at all, and are consequently marked by much irregularity in the choice of signs, and by a great waste of stenographic material. The object to be gained by a sound and complete analysis of signs is, of course, to assist shorthand inventors in selecting such signs and rules as will furnish the largest number of forms giving facile junctions, capable of being regularly and uniformly modified by the use of cross differentiations, and of such a character that their combination can never lead to ambiguity. I propose to divide this paper into two parts, the first an enumeration of all the possible ways in which one stroke may, for the purposes of writing, be made to differ from another, and the second an enunciation of the number of these differentiations which may be used together in any system.

The main differences between signs are seven in number; they are, as shown in Fig. V.

(i.) Direction.
(ii.) Hooks.
(iii.) Loops and Circles.
(iv.) Curvature.
(v.) Position.
(vi.) Size.
(vii.) Thickness.

Fig. V.

DIFFERENCES BETWEEN SIGNS.

Direction   Hooks   **Loops**   **Curvature**   Position   Size   Thickness

Motion   Motion of   Small   Normal   Double
of Arm   Fingers

                                  Thin   Thick

Right Handed   Left Handed

Initial   Final   **Initial**   **Final**

Straight Strokes   **Curved Strokes**

Simplex   Complex

Differing   Differing   Differing in   Variation
in Quantity   in Quality   of Quantity   of Quality

Positive   Negative

Lateral Position   Vertical Position

Thro'   **Against**   **Near**   **Far**   **Relative**   Actual

                 Superior Equal Inferior   Above On Through

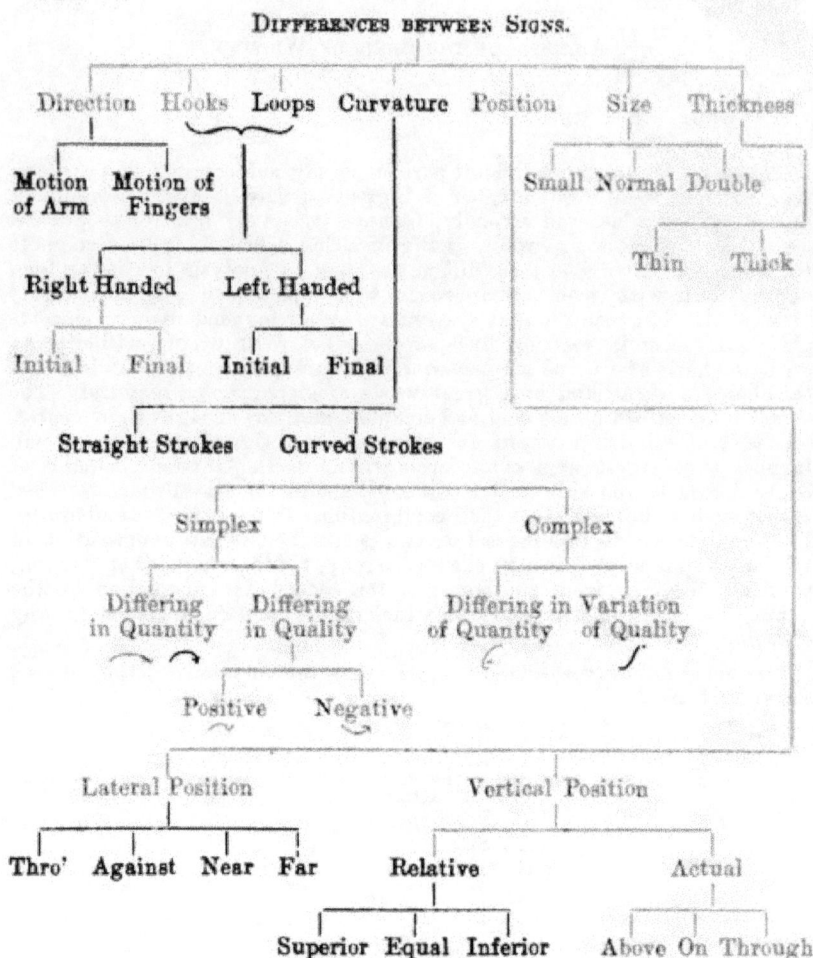

These we will proceed to analyse one by one.

(i.) The direction of a stroke is determined by two things, the motion of the arm, and the motion of the fingers. The **motion** of the arm is in the direction of **the** line of writing, the motion **of the** fingers is across it. When the arm **is at** rest, the easiest direction **in which** to write a stroke is that **in** which the fingers naturally move, namely, the slope / .

When the arm is moved forward, its forward motion is added to the motion of the fingers, and the easiest downstroke is now nearly upright, or even **from** left to right, according to the proportion between the speed of **the** arm and of the fingers, thus /\ ; the easiest upstroke is an oblique stroke from left to right, thus → . **Up and down strokes can express** more in a given time than horizontal strokes, because they allow the motion of the fingers to be superadded to the motion of the arm. The strokes ⟋ take no longer to write than the stroke ⸺ , but may be made to mean twice as much. The up and down strokes are those employed in Longhand and in the German Shorthands, but most English systems employ many more directions as distinct characters. The greatest number that can be safely distinguished is five, but some of these give rise to awkward joinings. The five directions used in some systems are :—

$$\swarrow \quad \downarrow \quad \searrow \quad \longleftarrow \quad \nearrow$$

(ii.) Hooks may be right or left-handed, and applied at the beginning or end of the stroke (see Fig. V.). The theoretical explanation of the different shapes of the characters in Longhand is that they are hooked characters joined by a certain rule that will be explained later on. Hooks are the most distinctive means of differentiation that we possess, but are liable in what are called geometrical systems to give rise to awkward joinings. The number of hooks which can be differentiated, as applied to straight strokes, is four ; when applied to a curve hooks can only be written on the inside, which reduces the number of differences to two.

(iii.) Loops and circles may also be initial or final, and right or left-handed. It is almost impossible to distinguish circles from loops in the middle of words, and difficult to do so at the beginning or end.

(iv.) Curvature distinguishes, firstly, straight strokes from curves ; secondly, simplex from complex curves. Simplex curves are those whose curvature is homogeneous throughout, that is to say those which are segments of circles ; they differ from one another in two ways, in quantity and in quality (see Fig. V.). The quantity of a curve varies inversely as the length of the radius of the circle of which it is a segment, that is to say that the quantity of curvature in a segment of a small circle is greater than in a segment of a larger circle. The strokes ⌣ and ⌢ differ from one another only in the quantity of their curvature. The following fact with regard to quantity of curvature seems obvious, but has often been disregarded. If half-circles are used in any system of writing, then no quarter-circles of the same quantity of curvature must be employed, for fear lest coming together they might produce ambiguity. But apart from this, mere difference of quantity of curvature is insufficient to differentiate two strokes for practical Shorthand purposes. The difference between ⌒ and ⌒ , signifying in "Phonography" mtr and lr, practically disappears in actual writing. To pass to quality of curvature, it is sufficient to notice that in quality a curve may be either positive or negative, positive if drawn in the same direction as the hands of a watch, negative if drawn in the opposite sense. Finally, complex curves (see Fig. V.) are those whose curvature varies. The variation may be of two kinds. Just as simplex curves differed in quantity and quality, so complex curves differ by *variation* of quantity and quality, thus :— ⌒⌒ and ⌒ differ in variation of quantity, and ⌒ and ⌒ differ in variation of quality. Complex curves cannot be used to any considerable extent in Shorthand, because they are liable to cause confusion with combinations of simplex curves.

(v.) Position may be of many kinds. It may in the first place be lateral or vertical (see Fig. V.); lateral position means horizontal distance from the preceding character. Under the most favourable circumstances, four grades of lateral position may be distinguished; the second character may be written through, against, near to, or far from the preceding, thus:—

┼ ; ┝ ; ┝— ; / — . Of these, however, only the last two are generally applicable, and any use of the others necessarily involves a number of special rules and consequent irregularity. The most interesting and skilful use of lateral position is that found in the Arabic numerals, but I cannot at present see any means of applying to Short-hand rules similar to the rules of numeration. The most elaborate application of lateral position to Shorthand purposes is that of Professor Everett, but he carries the device too far, and the result is a good deal of difficulty and irregularity. Vertical position may be relative, that is with reference to the preceding character, or actual, that is with reference to the line of writing. In relative vertical position three grades may be distinguished, superior, equal, and inferior, thus // ; // ; and // . It may also be *regulated* in various ways, *i.e.*, the relation may be between the ends of the respective characters, or between their centres, or between their commencements. Thus in the pair // , the second character would in the first case be in superior relative position as compared with the first; in the second case it would be said to be in equal position; and in the third, in inferior position. Leaving relative vertical position, we come to actual vertical position (see always Fig. V.), or position relative to the line of writing. Three grades are generally distinguished, above, on, and through, thus /, /, /. There is a French system by L. Chauvin, published in 1853, in which, among other features of the very greatest interest, is a means of indicating four positions by the help of what is called "la barre," a line running parallel to the line of writing at the distance of a short consonant. As a matter of fact nothing is practically gained by this device, but it is ingenious and well worth recording. Other French systems have made a much more extravagant use of position, proposing to write on the five lines of the musical stave. The following is a specimen from the French system of Astier, which I have copied from Mr. Anderson's famous and invaluable History of Shorthand.

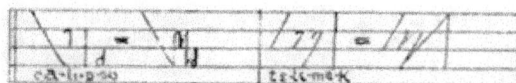

The words represented are *Calypso* and *Télémaque*. In Longhand four positions are observed, but position is never the sole distinction between signs. One of the peculiar features of Longhand, which will be fully discussed presently, is that very few characters differ from one another in less than two ways. Finally, strokes in all the positions enumerated may in theory be written either disjoined or joined. This, however, is only practically possible when many of the differences of *direction* are sacrificed to the principle of the connecting upstroke.

(vi.) Size has apparently been considered by many Shorthand inventors to be a most distinctive means of differentiation. It is, however, quite the reverse unless kept within narrow limits, inasmuch as it is not, like direction, hooks, loops, and some kinds of position, an actual difference, but only a relative, like the next difference, that of thickness. Probably three sizes may be safely distinguished, the small size for hooks and circles, like those in the Longhand letters, ι , λ , and ℓ ; the normal size, like the Longhand o ; and the double size, like the Longhand ℓ .

In Longhand, it may surprise some persons to learn, six sizes are used, but size, like other differences, is but rarely used in Longhand as the sole difference between strokes.

(vii.) Thickening is the least trustworthy of the means of differentiation, and as often used is certainly as great a detriment to facility as it is gain to brevity. The obvious restriction which should be observed is this: it can only be advantageously applied to strokes nearly in the direction of the slope of the pen. There are two ways in which the pen can be held, either pointing in a direction nearly at right angles to the line of advance, in which case only downstrokes should be thickened, or else pointing in a direction nearly parallel with the line of advance (the position often used by Shorthand writers), in which case only strokes from left to right can be conveniently thickened.

We have now considered all the possible differences between strokes. I pass now to the second part of this paper, and shall examine how many of them may be used together in one system. A complete investigation of this question would be almost equivalent to the invention and description of an infinite number of Shorthand systems. It will be sufficient to show that there are certain main types of writing systems, and that, for any given type the modes of differentiation which can be used without ambiguity can with some accuracy be calculated.

The chief difference between systems is this. The characters may either be joined directly one to the other, the second beginning where the first ends, or they may be placed side by side on or near the line of writing, and connected together by upstrokes having no inherent meaning. I have ventured to call these two main types, for which no satisfactory names have as yet been proposed, the Articulate, or directly-jointed systems, and the Linear or Alternating systems; and for the upstroke I adopt the French word Liaison. In Articulate systems, to which class most of the English Shorthand systems belong, the same scheme of differentiations is applied to strokes in all directions. In the Alternating systems, to which belong most of the German Shorthands, some English systems, and Longhand, strokes must be distinguished as upstrokes and downstrokes. The latter have a certain scheme of differentiations applied to them, the former are as a rule not differentiated, though there is no good reason why they should not be differentiated by curvature, as they are, successfully, in "Linear Shorthand."

It can hardly be necessary to enlarge on the fact that the liaisons cannot be differentiated in the same way as the downstrokes or characters. I may, however, just point out, first, that the function of the liaison being to connect two characters, which must be in certain relative positions, its two extremities are necessarily fixed points, and it cannot therefore be differentiated by length or direction ; secondly, that the hooks and circles at the point of junction between the liaisons and characters being used as a means of differentiation of the latter, we cannot employ them to differentiate the former. In Linear or Alternating systems then we have an *alternation* of characters and liaisons, significant and non-significant strokes ; in Articulate systems we have a simple *articulation* of each character with the extremity of the preceding. The primary object of the Linear principle is to place the characters in a convenient position and so assist facility, but it is to be particularly noticed that we incidentally, as it were, obtain a very important addition to the scheme of differentiation of the *characters* by the possibility of joined position. Instead of placing the second character on the line, beside the first, we may place it in any one of six positions, thus multiplying its signification by six without interfering with the lineality of the system.

Having described briefly the characteristics of the two main types of systems, Articulate and Linear, I proceed to notice various subdivisions of them, and will commence with the Articulate type as being the oldest.

Fig. VI.

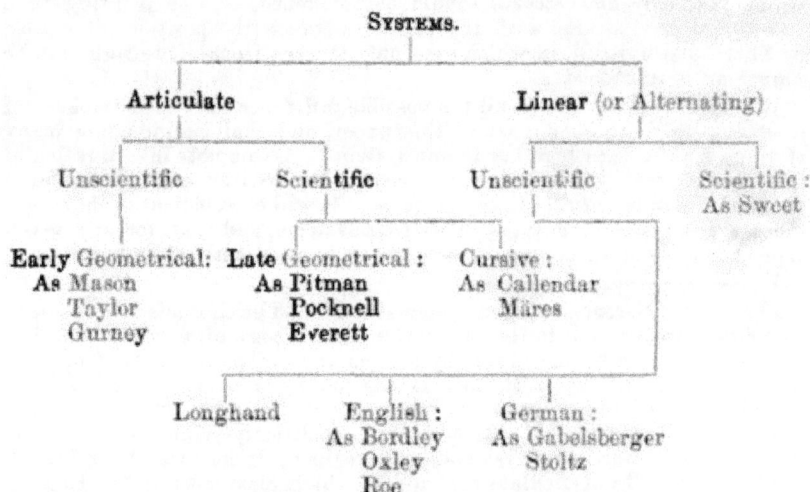

Articulate systems may be divided into two classes, scientific and unscientific. The only idea of the inventors of the unscientific systems was to obtain a sufficient number of brief and distinct signs to represent the letters of the ordinary alphabet and a few common combinations. The aim of the inventors of scientific systems is to arrange the alphabet in such a way as to use completely all the differentiations employed, and to be able to modify their alphabetic characters in various ways by comprehensive and uniform rules. With this object in view, it is almost necessary that the Shorthand inventor should arrange his alphabet on Phonetic lines, since the modifications required by a *class* of allied sounds are more regularly and satisfactorily provided for when those sounds are represented by a *class* of allied signs.

I.—The first type that we have to consider is therefore the "unscientific articulate" type. These systems, the only systems known in modern times until the publication of "Phonography," were, as a rule, founded upon an elementary geometrical analysis of signs, and may conveniently be called the "early geometrical systems." The inventors, in order to obtain the simple strokes to which their ambitions were limited, took first the straight line, as the simplest stroke of all, and used it in five directions. This gave five characters, $/$, $|$, $\backslash$, $—$, and $\diagup$. Next they took two curves in the direction of each straight line, often, however, rejecting the "oblique" curves. This gave from four to ten characters more. The addition of a circle at the beginning of each stroke, to be used on either side at will, raised the total number to from eighteen to thirty. Finally they applied a hook to a few strokes and this gave them the requisite number of characters.

This is the simplest form of system. Its distinguishing features are, length, simplicity, awkwardness, distinctiveness, and lack of arrangement. It may safely be said that the old geometrical systems can only be used for reporting after very long apprenticeship, and by omitting a considerable part of both words and phrases. The means of differentiation used in these systems are four in number, direction, curvature. hooks, and circles. The last two are not worked out to their fullest extent, and it is easy to show why this is so. The number of signs so added would not be sufficient to change the character of the system; they could only be applied to *independently selected combinations of letters*, and so would introduce complication without a proportionate increase of speed. Any system, therefore, confining itself to the use of the above four means of differentiation of signs will probably use them to the same extent and in a similar manner, and will certainly have the character of the old geometrical systems.

II.—The scientific articulate systems may be divided into geometrical and cursive. The first class (which may be called late geometrical in contradistinction to the early geometrical unscientific systems), differs from the second class, or cursive, by the fact that the former considered brevity to be equivalent to speed, while the latter recognize (a fact that I hope to bring out fully in my next Part) that speed depends upon two things, brevity and facility. With a view to obtaining the utmost brevity the inventors of the late geometrical systems made use of the largest possible number of differentiations, rightly considering that the brevity of a system varies directly as the number of differentiations employed. The inventors of the cursive type, with a view to facility, select chiefly such forms as are on the slope of ordinary writing.

We will take first the geometrical systems (see Fig. VI.). The alphabets of these are based upon a geometrical analysis like the early geometrical type, but in addition to an alphabet much wider than those used in these last, and arranged in a methodical and orderly manner, they provide various means of *modifying* the alphabetic characters. They may all be divided into two parts, enumeration and modification, of which the first only is found in the early geometrical methods. They are characterised by brevity, want of facility, orderliness, lack of simplicity, and lack of distinctiveness, and are suited best to professional reporting. The means of differentiation of signs used are more numerous than in any other type, and include all those enumerated in the first part of this paper, except complex curvature and some kinds of position. It is theoretically impossible to use all these differentiations in one system, and in practice their use leads of course to the existence of numerous "incompatibilities," which are one of the distinguishing features of the type under discussion. These incompatibilities are either reconciled by a multitude of detached rules, or are left as permanent sources of confusion. It may be well to notice one or two of these in detail to illustrate the point.

(i.) Where differentiation by size is used it is theoretically impossible to repeat a straight consonant. This fact is always disregarded. In Pitman's phonography for instance, _____ might mean **k**, or **kt-kt**, if it were not regulated by a special clause. One case is actually left unprovided for. A double length straight letter ending in a hook has two distinct meanings, ——⟶ = **k-kn** or **k-ntr**. That is, **kn** and **ntr** are represented in the same way. I believe that in the ordinary form of phonography *pippin* and *painter* are represented by the same outline.

In Professor Everett's system the same difficulty of course occurs, and is evaded by special rules as to the use of a hook. The same is the case in Mr. Pocknell's Legible Shorthand, while in one of Melville-Bell's systems the second character is written in position at any convenient angle. The only way of avoiding this incompatibility in an articulate system is, it seems to me, to represent the consonants by curves, and the vowels by straight lines. In English a vowel is never sounded twice consecutively.

(ii.) The second incompatibility which we will notice is that caused by a too full use of disjoined position, as in Everett's system. Some characters for instance cannot be written across some others, and so special rules are rendered necessary.

(iii.) Again, it is impossible to use all the four kinds of hooks when all directions are used. Back hooks in the middle of words constantly present impossible joinings. Thus in phonography *subskibr* is written for *subscribe*, because the r-hook cannot be formed.

(iv.) Fourthly, the use of hooks clashes with that of curves. Entirely apart from the fact (which will be established presently) that it is well not to distinguish between curves and hooked straight lines, the impossibility of putting hooks on the outside of curves produces exceptions which are confusing. Thus in Pitman's system l after **ch** is represented by *a small left-handed hook* (⌒ = **chl**), but after **sh** (the corresponding "continuant"), it is represented by a *large right-handed hook* (⌒ = **shl**).

Such is the nature of the scientific geometrical type. Any system which aims at the greatest possible brevity by using the greatest possible number of differentiations will be a system of this type, and will use these means of differentiation in a similar manner.

Passing to the cursive type we find a marked difference in character and appearance. The leading idea of the inventors of cursive systems is to secure greater *facility* than is found in the geometrical. In forming the alphabet the first step is to *construct* a geometrical analysis, the second to *reject* such signs as present awkward joinings, and the third to *select* rather irregularly certain other signs (chiefly complex curves), which a geometrical analysis will not provide. The differentiations used are less numerous than in the preceding type. They are: direction, length, curvature, circles, position, hooks (named in the order of their importance). None of them are used to their full extent. The principle of *rejection* that prevails in these systems reduces the directions to four, lengths to two, confines curvature to certain directions, allows a fairly wide use of disjoined position and circles, but makes little use of hooks. Differentiation by thickness is generally altogether rejected, but might be applied (as indicated in the early part of this paper) to strokes drawn in the direction of the slope of the pen. The distinguishing features of the cursive type are, general facility, with occasionally great infacility of junctions, legibility, irregularity, and want of brevity. Any system formed with a view to facility on articulate lines, will be a cursive system, and will probably use the means of differentiation in a similar manner to that indicated above. It will not be able to make much use of the fifth direction, or of thickening, because they take away from the facile character desired; nor of hooks, because in articulate systems they cause awkward joinings; nor of joined position, because upstrokes are used as characters. The leading system of this type is Callendar's "Cursive Shorthand," which is certainly one of the best systems published for general use. Its chief faults are its length and the number and irregularity of its alternative characters.

It is a curious fact that some inventors of cursive systems have made a return to the Orthographic principle. The advantages of Phonetic arrangement are orderliness and brevity: its only disadvantage is its novelty, and it is difficult to see why a return should now be made to the older method. Orthographic cursive systems, of course, cannot provide those means of regular modification which give the scientific character, and therefore should be included in the unscientific type in a class parallel to the early geometrical. I have not noticed them in that place, because they are not sufficiently numerous or important to demand classification under a separate head.

We now come to the Linear Systems, whose main feature is, as I have said, that the characters alternate with liaisons. They may be divided, like the Articulate systems, into scientific and unscientific, the first aiming only at a representation of the individual letters, the latter at an alphabet arranged on scientific lines, and at the comprehensive rules which such an arrangement renders possible.

III.—The unscientific Linear systems may be divided into three classes, containing Longhand, a few old English systems, and the modern German systems. I will take first the old English systems. Of these the chief are Bordley's, Oxley's, and Roe's. The means of differentiation used are: (i.) For downstrokes: hooks, joined position, and size. No distinction of direction is made, and the differentiation by circles or loops is not used as distinct from that by hooks. This latter point is an important one, and will be considered presently. (ii.) The upstrokes or liaisons are not differentiated. (iii.) An anomalous significant horizontal stroke is sometimes used.

The features of these systems are their resemblance to Longhand, their facility, and their length. The only one which is really familiar to me is that of Roe, which is far too long for reporting as presented, while the unscientific nature of its foundation would cause any abbreviating rules to lead to confusion. The type is settled by the use of only one direction and of back hooks. The first limits the number of characters to such an extent that position has to be used up in providing characters, and the use of back hooks prevents that of loops, so that there are no materials left for abbreviating rules. Differentiation by thickness might and should be employed. Any Linear system using back hooks and only one direction would be of the type and character of the old English Linear systems.

The German systems are suggested by, if not derived from, Gabelsberger's. That great inventor composed his alphabet without any scientific basis, and consequently his system, and the other German systems, have the non-scientific character which is found in, *e.g.*, Taylor and Gurney. They are not based upon any analysis, however imperfect, and consequently the signs are chosen at haphazard, many good ones are not selected, and many dangerous ones are used.

The differentiations used are: (i.) For downstrokes: position, hooks, length, direction, curvature, and thickness. (ii.) Upstrokes are, of course, mostly liaisons which are not differentiated, but there are several instances of what I may call *anomalous upstrokes*, which are not liaisons, and which interfere very seriously with the regularity and lineality of the systems, preventing the universal application of principles. This is one of the incompatibilities in German systems. It is found in upward t and f in various systems. These letters, whenever they occur, disarrange the line of writing and necessitate a special rule, and in some cases a special kind of joining. Another of the incompatibilities which are somewhat numerous in the German systems is caused by the use made of thickening to indicate vowels. It is sometimes the preceding and sometimes the following character that is thickened, and cases are bound to occur where a character thickened to modify the preceding vowel may be read as modifying the following.*

The distinguishing features of German methods are facility, irregularity, and brevity. They are used with great success, both for reporting and for correspondence, and enter far more into general use than any English system. Gabelsberger's and Stoltz's systems have been admirably adapted to English. The type of the German systems is settled chiefly by the haphazard selection of the alphabet. Any alternating system having an alphabet of similarly selected characters, would naturally adopt a similar use of the differentiations of signs. The haphazard selection of the characters prevents the complete use of differentiation by direction, curvature, and hooks (the three most important kinds), and therefore there are not sufficient means left after the construction of the alphabet for regular modification of the characters.

Longhand is, of course, much more difficult to analyse than any other system, and can only be approached on the lines on which we have been working. The signs were not designed as at present existing, but have been gradually evolved by a slow development from Uncial Capitals. They are, therefore, far more unscientifically arranged than those of any other system, and by no means the best that might be obtained with the same means of differentiation. The means of differentiation found in longhand are three in number: hooks, joined vertical position, and length or size. The liaisons are not differentiated, but there exist several cases of anomalous upstrokes in the latter half of letters, as for instance in *c*, *l*, and *w*. They are joined by a short horizontal curve, blending with the liaison. Perhaps it would be more correct to consider these cases as modifications of characters, by curvature

* This subject has been already partly discussed in the Preface.

of the following liaisons, thus *bl* differs from *dl* by the form of the upstroke.

The differentiations of size and position are used in a peculiar manner, which rather conceals their nature. They are combined, as shown in Fig. VII., seven different kinds of strokes being used, whose length and position are defined by the points where they begin and end.

Fig. VII.

Defining Points.	On the Line.	One Unit above.	1½ above.	2 above.
On the Line		*l. a.*	*l. c.*	*l. l.*
1½ Unit below	*z*	*j. g.*	*j. p.*	*f. f.*

The points used are five in number. On the line, one unit's length, one-and-a-half units' length, and two units' length above the line : and one-and-a-half units' length below the line. These give the seven different kinds of strokes that are observed in correctly written longhand. They are of six different lengths. These differences are not sufficiently distinctive to be used alone, and we shall immediately see that it is one of the peculiarities of longhand that no two characters differ in less than two ways. The only other differentiation used is, as has been said, that by hooks. All the possible hooks are used, initial and final, right and left handed, but the use of them is disguised by a rule that I have alluded to before, and that it is now time to examine.

There are three most important and peculiar principles to be found in the arrangement of longhand, considered from a theoretical point of view, the first two of which are deserving of special attention. They are as follows :—(i.) Curvature follows the hooks ; (ii.) back hooks become circles when joined ; and (iii.) no character differs from another in less than two ways.

(i.) Curvature follows the hooks. When a hook is placed on a straight line, it invariably tends to curve it. This is well known to all shorthand writers. In longhand no distinction is made between hooked straight lines and curves ; in other words, the characters may be looked upon as hooked straight lines, which may be curved as convenient to make the outline more facile. In nearly all shorthand systems the character *C* would be distinguished from *(* ; ** from *)* ; *⁄* from *∫* , &c. In longhand they are not so distinguished, and consequently it may be recklessly scribbled without obscuring the meaning. Of course such a rule involves a considerable waste of material, but this is more than compensated for, I think, by the gain in elasticity and facility. This principle has been employed in Linear Shorthand.

(ii.) Back hooks become circles when joined. In most systems a good deal of ambiguity, and much irregularity, is introduced by the use of back hooks and forward circles, like *∫* , *⁄* , *ſ* , and *6* . In longhand only * forward hooks and backward circles are employed, like *l* , *⁄* ,

---

* The only exception is the letter *c* , which has to be written disjoined.

c , and *y* , except at the end of words. Thus *ʃ* at the end of a word terminates in a back hook but when followed by another letter this becomes a back circle, thus *ʃ* The use of this principle involves either the surrender of the differentiations between circles and hooks or the disuse of back hooks altogether. The circle is, of all others, the device best adapted for the comprehensive abbreviating rules of which I have spoken as being a necessity in a scientific shorthand, and it is not, therefore, difficult to decide that if this principle is adopted, as it certainly should be in every alternating system (and is in Linear Shorthand), the use of back hooks must be abandoned.

(iii.) No character differs from another in less than two ways. This is one of the most striking features of Longhand, and one which gives the writing a very *distinctive* character, but it is not necessary, provided that the characters are selected upon proper principles, and, of course, it reduces the number available by more than half. Consequently, in Longhand, most of the letters are formed by a combination of two strokes, joined by the liaison. The total number of simple strokes to be obtained by the differentiations used in Longhand is twenty-eight, which is more than sufficient to provide a complete alphabet of simple strokes. The rule which we are considering, however, that each character must differ from every other in at least two respects, halves this number, and makes it necessary that two strokes, roughly speaking, should be used to represent each letter.

The distinguishing features of Longhand are distinctiveness, facility, and length. Its type is settled by the three rules just examined, and any alternating system observing these rules will be forced to use the same number of differentiations, and to use them much in the same way. Of course it would not be possible to construct a practical Shorthand upon these lines.

IV.—We now come to the last type of system, the Scientific Linear. The only instance of this type, so far as I know, is the system of Sweet, which is of great interest, and should be in the hands of every student of Shorthand science. Though the sounds are well arranged for representation, the use made of the differentiations admitted does not give nearly the best possible results. Position and size together are only made to give four strokes, although a quadruple length is used, as in Longhand. Direction and thickness are unnecessarily rejected, and loops are used irregularly. The system is legible, facile, and distinctive, but partakes too much of the nature of Longhand to be a successful Shorthand. As no system has, as yet, been published on scientific alternating lines, which takes full advantage of the possibilities of this type, it is impossible to say that such a system could be made to possess such and such characteristics. Both the scientific character of the late geometrical systems, and the alternating character of the German systems, present certain advantages, and the question which presents itself is whether the combination of their characteristics would provide a better system than either, and whether it is actually possible to combine them. These questions I shall attempt to answer in my next, and final, paper.

PART III.

COMBINATION OF SOUND AND SIGN.

Having considered pretty fully the sounds which have to be represented in Shorthand, and the signs that are available for that purpose, I am now able to examine how these may best be combined into a Shorthand system. And first we must decide which of the various types explained in Part II. is the most suitable; and secondly, discover the best way of assigning the various signs which the selected type provides, to the sounds which require representation.

The rational way of deciding the type is shown in the form of a diagram in Fig. VIII., to which I shall refer from time to time as I proceed

FIG. VIII.

INVESTIGATION OF THE MOST SUITABLE TYPE.

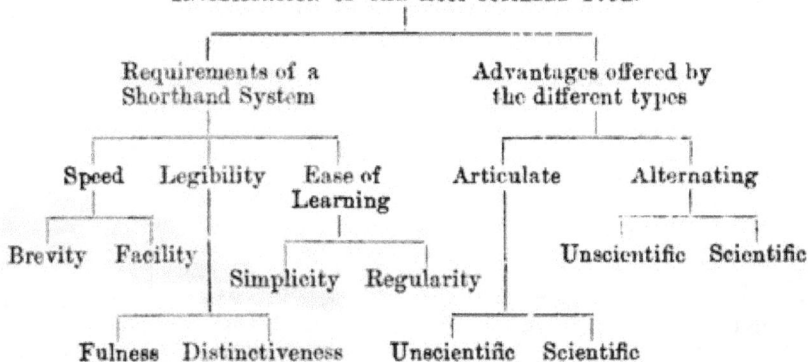

We must first see what are the requirements of a system, and then compare with these the facilities offered by the various types.

It is of course obvious that the main requirements of a Shorthand system are that it should be fast, legible, and easy to learn. Of these the first is perhaps the most important, and the last the least so.

Speed, as indicated in Fig. VIII., depends upon two things, brevity and facility. Brevity has often been taken to be equivalent to speed, but it is far from being a fact that the shortest outline is the quickest. The strokes ⌄ are longer than the stroke ⌄ , but can be written much more rapidly. Facility is indeed quite as important a feature as brevity. We must not only therefore look for the largest possible number of simple signs, but must be careful that they are facile as well as brief. As shown on page 10 the strokes ⌐ and ⌐ are the most easily formed, and should therefore predominate. But more than this is necessary; there must be some provision for ensuring that the joinings between the characters are facile. This is always a great stumbling block to inventors, who have generally had to provide alternative characters for many sounds, in the hope that one may provide a facile joining when the other does not. This plan is never satisfactory, because it entails a good deal of uncertainty in some words, as well as a considerable waste of material. In fact the only way of absolutely insuring the impossibility of a bad joining, seems to be to adopt the principle of the connecting up-stroke, used in Longhand and the other alternating systems. Speed, therefore, involves the necessity not only of a wide alphabet of simple and facile strokes, with the directions ⌐ and ⌐ predominating, but also some provision for ensuring facile junctions.

Legibility in a similar way may be resolved into two elements, fulness and distinctiveness. It is of course obvious that the omission of sounds is detrimental to legibility, and that the fuller the expression of a word is the more easily it is read. The necessity of brevity, however, compels us to reduce to the lowest limits the number of signs used in a word, and it is therefore necessary to examine what are really the important sounds used in language, and how they can be most concisely reproduced. Ever since the first modern system was published, the principle of representing the sounds and not the letters of a word, has been more or less recognised, and gradually new phonetic characters were introduced to take the place of combinations used in the ordinary spelling. Such combinations for instance as *ough*, *eye*, *aw*, and *th*, cannot be represented in Shorthand by their alphabetic characters as spelt, and therefore, if for no other reason, it seems necessary that a modern system should be founded upon a phonetic basis. Fulness of expression also requires that vowels should be expressed as well as consonants, and that they should be provided with means of distinct representation in the body of the words. The necessity again of having signs that may be *distinct* from one another, both when standing alone, and also when connected in any sequence, requires that the signs to be employed, as well as the sound to be represented, should be carefully analysed beforehand.

Ease of learning is also an important feature of a Shorthand system, and it depends upon two things, as indicated in Fig. VIII., simplicity and regularity. Most inventors, especially the earlier ones, have laid too much stress upon the first of these and too little upon the second. The saying that what is learned without trouble is generally useless when learned, applies very forcibly to Shorthand. A mere alphabet of simple strokes, such as was provided by inventors up to the time of Pitman's Phonography, is not only less useful than a more complicated system which provides regular means of abbreviation, but is actually more difficult to learn, in spite of its simplicity, because of the irregularity which becomes necessary when it is to be written at high speed. Other things being equal, however, the simpler a system is the more easily will it be learned, and therefore it is important that the alphabet should be arranged in groups of allied signs representing groups of allied sounds. Regularity, however, is a far more important attribute of a Shorthand system. The great aim of the inventor should be to arrange the characters in such a way as to make them amenable to certain great rules of abbreviation, which may be uniformly applicable throughout the system, and the principle which may be considered to be now pretty well established is that the small circles, loops, and hooks, should be devoted to this purpose, and governed by a few symmetrical and comprehensive rules. It is this principle, more than any other, that makes it imperative that the alphabet should be arranged in groups or classes of allied signs representing groups or classes of allied sounds. Obviously certain modifications are often required by one group of sounds which are seldom or never required by another, and hence it is important that the signs representing a group should all be equally capable of conforming to some device, which need not necessarily apply to another group. Thus, vowels should be represented in an entirely different way from consonants, and short vowels in a different way from long vowels. Dental consonants should all have similar forms, while labials have forms of another character, and so on.

Let us briefly resume the principles which we have now evolved from the primary requirements of speed, legibility, and ease of learning. They are as follows :—

(i.) We must discover as many simple strokes as possible, our choice being limited by the principle that they must be facile and distinct, the directions ╱ and ╱ predominating.

(ii.) Some effective provision must be made for ensuring invariably facile junctions.

(iii.) The alphabet must be scientifically based upon analyses of signs and sounds. Both consonants and vowels must be provided for, and must be arranged in groups of allied signs representing allied sounds.

(iv.) Provision must be made for the regular and comprehensive use of symbols as modifiers.

Let us now compare these requirements with the advantages offered by the different types of system, as detailed in Part II.

I.—The unscientific articulate type may be quickly dismissed. Almost the only advantage presented by such systems over the scientific type is their simplicity, and this, as has been said, is often a disadvantage in the end.

II.—Of the scientific articulate type we will first compare the geometrical systems, and then the cursive. The former, as shown on page 15, are characterised by :—(i.) brevity ; (ii.) lack of facility ; (iii.) orderliness ; and (iv.) complication. Therefore our first requirement, of a large number of signs, is well met, as are also the third and fourth requirements of an orderly arrangement of alphabet and devices. The principle, however, that the signs must be facile and distinct is but ill provided for, and the requirement as to invariably facile junctions must be altogether neglected in a system founded upon these lines. The Cursive type is distinguished (i.) by general facility of characters, (ii.) with occasional great infacilities of junction ; (iii.) by legibility ; and (iv.) by want of brevity. It therefore conforms well to our first requirement, of facile and distinctive signs, and to the third requirement as to scientific tabulation.

Owing, however, to its hybrid character, it affords but poor facilities for the regular use of symbols as abbreviating devices, and it shows curious lapses from the principle of facile junctions. In fact, as has already been suggested, the only means of absolutely preventing the occurrence of occasional awkward joinings, is to adopt the principle of the alternating systems, the connecting upstroke. I proceed, therefore, to examine whether they adequately fulfil the other requirements which we have laid down.

III.—Of the unscientific alternating type, it will be sufficient to consider the German systems, as they possess most of the advantages of Longhand and the early English alternating systems, and other advantages besides. The German systems are characterised as was shown on page 18, by (i.) facility, (ii.) irregularity, and (iii.) brevity. They conform very well to both parts of our first requirement, and to the second regarding junctions, but fail with regard to the two last requirements as to scientific tabulation of the alphabet and uniform abbreviating rules. It is in these respects that

IV.—The scientific alternating might be made superior to the German type of system. It would have the same immunity as the latter from awkward signs and junctions, it would have the same numerous and facile characters, distinct whether alone or in connection with others, while it would combine with these advantages those which are possessed by the scientific articulate systems, the orderly scientific arrangement and the uniform and comprehensive rules of abbreviation. It is this type, therefore, which, from what we have shown, seems to afford the best opportunities for constructing a practical shorthand system.

I now come to the second part of this paper, the assignment of sign to sound, and, as in the first part, will give a diagram of the steps which we shall follow.

Fig. IX.

THE COMBINATION OF SOUND AND SIGN.

Tabulation of Sounds.	Tabulation of Signs.	Assignment of Sign to Sound.
Vowels. Consonants. Combinations.		The two Rules. The Compromise.

Upstrokes. Downstrokes Symbols.

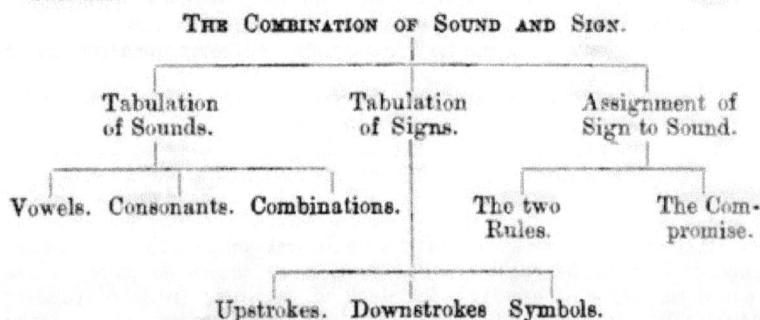

We must first tabulate the sounds which we have to represent, and for this purpose must refer again to Figs. II. and IV. (pages 4 and 7). The vowels, as we have seen, consist of eight simple and sixteen compound vowels. It will be remembered, however, that the short vowel aw, as in *salt*, is only found before l followed by a surd, and that in this position ă, as in *sat*, is never found, so that it will be sufficient if we represent both these sounds by one sign. Again, the prolonged sound of aw, as in *sword*, is very common, while the prolonged sound of ă is not used, so that aw may be represented by the long form for **ă**, **aw**. The short vowel **o**, as in *whole*, is only used by a few persons, and may be omitted. The sounds **ah** and **ur**, as was said, though really compounds of the **ŭ** termination with **ŏ** and **ŭ**, are so nearly simple prolongations of these sounds that they may be tabulated as if they were so. The remainder of the compounds formed with the **ŭ** termination are best left for a separate rule, and it will thus be seen that each simple vowel has one compound or long vowel attached to it, and that there are besides four compound vowels whose bases are not used, thus :—

Simple Vowel.	Long Vowel.
ă	aw
ĕ	ay
ĭ	ē
ŏ	ah
ŭ	ur
oo	oo

and
oy, ō, î, ow

Turning to the consonants, and referring to Fig. IV. page 7, we find that there are twenty-four used in English. Of these we may omit **wh**, **zh**, **kh**, and **gh**, on account of their rarity, so that the consonants for which we shall have to provide signs will be :—

P B	F V	M	W			
T D	Th Dh	N		S Z	R	L
			Y	Sh		
K G		Ng				

and the aspirate H

Lastly, we must see what sounds must be provided with special facilities for combination with others. We have already seen that the termination **ŭ** is to be so treated, and an examination of the consonant groups, which are so common in English, will show us that the following sounds require to be specially provided for :—

s, either as the first or second in a combination.

n, generally as the first in a combination.

r and **l**, as the second : generally at the beginning of words.

w and **y**, as the second ; generally in the middle of words.

t, as the second ; generally at the end of words.

Having tabulated the sounds required, we must now tabulate the signs available. Taking it for granted that we adopt the Scientific Linear or Alternating type of system, and the rules that curvature follows hooks, and that back hooks become circles when joined, as we have already decided to do, we have to consider what forms the differentiations allowed by this type can be made to give, and which of them will be best used as characters and which as modifiers. And first, it seems obvious that we should represent the consonants by downstrokes ; the vowels by upstrokes ; and the modifiers by the small hooks and circles, which are neither upstrokes nor downstrokes, but may be considered theoretically as horizontal. This arrangement is a very suitable one for many reasons, and I have adopted it in Linear Shorthand.

The upstrokes, according to the principles of Alternating systems, are not differentiated in the same way as the downstrokes. They are used to place the latter in position. In Longhand and in the early English Alternating systems they were only used to put the characters in position side by side on the line. It was Gabelsberger who first saw that they might be used to indicate the vowels by putting the consonants in various positions with regard to the line and to each other. In themselves, however, they cannot put the consonants into more than six different positions, and this is not enough to distinguish between all the vowels required, and we are obliged to use a cross differentiation. The German authors have used thickening for this purpose, not noticing that it is easy to differentiate the liaisons themselves by *curvature* and reserve thickening for doubling the alphabet. By distinguishing between straight and curved liaisons we increase the number of differentiations to twelve, and if we admit "double" curvature we add another six, making eighteen in all, or two more than we require. Our vowel scheme will therefore be as follows :—

LIAISONS.	LATERAL POSITION.		VERTICAL POSITIONS.
	NEAR.	FAR.	
Straight.	above on below	above on below	
Curved.	above on below	above on below	
Double curved.	above on below	above on below	

Coming next to the Downstrokes or Consonant signs, I shall take the various differentiations, as shown in Fig. V., page 10, and multiply them one into the other to find how many distinct characters we have at our disposal. Beginning, therefore, with Direction we find three strokes available, / , | , and \ . Taking / first as being the most facile, and multiplying it by the differentiation of hooks, we get the series :—

Of these the last five contain back hooks, and we are strictly bound to reject them because they interfere with the use of the circle, under the rule that back hooks become circles when joined. Two of them, however, ⟮ and ⟯ , are so clear and facile that we can hardly afford to waste them, although they will introduce a certain amount of irregularity. The former is very convenient at the commencement of words, and one is tempted to use it for c (**k**) but as it is an anomaly in a system on the lines indicated (as it is in Longhand), and as we cannot obey the law by connecting it with a circle or loop, because we are using loops in another way, we are forced to use it disjoined except at the commencement of words, and it is therefore best to assign it to a sound that generally occurs in such a position, such as a combination with **r**. The back hooked character ╱ cannot be written in connection without introducing circles, as ⨍ , and on the other hand it would be extremely awkward to write it disjoined, because there would have to be a hiatus both before and after it. In spite, therefore, of its facility and distinctiveness it would be better not to use it unless it could be assigned to a sound with which the sound represented by the circle never combines. In such a case a non-significant circle might be introduced, though it would be an undesirable exception to the general rule.

So much for the differentiation by hooks. The differentiation by loops and circles is reserved for the purpose of providing modifying symbols. The differentiation by curvature is excluded by the rule that curvature follows hooks. The differentiation by position has been already dealt with. The differentiation by size and thickness is deferred for a moment while we return to the second direction ╱ . This cannot in practice be distinguished from the third direction ╲ , except by writing the characters slowly, and we, therefore, reject the least facile of the two, namely, the former. The third direction, like the first, may be differentiated by hooks, giving the characters,

Of these we reject the four back hooks ⟨shorthand characters⟩ ; the stroke ╲ cannot be easily joined if written straight, it is therefore curved thus ╲ , almost like the back hook ╲ , and is then introduced with ease in any combination. The strokes ╲ and ╲╲ (which become ╰ and ╮ in practice, in accordance with the rule that curvature follows hooks), are not so facile as some of the other strokes, but do not present any difficulty whatever, and are very suitable at the extremities of words. The stroke ╲ is one of the most facile in the alphabet, and should be assigned to a common sound.

All of the signs thus arrived at may be differentiated by size and thickness, which will give us four distinct signs for every one of those above, making our total number of simple downstrokes up to the total of forty, which will be found sufficient, not only for the simple consonants, but for a number of combinations as well. Thus :—

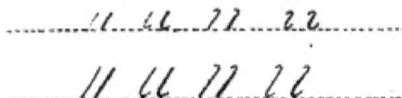

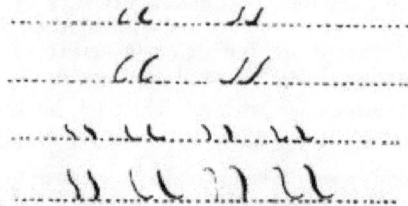

Finally, we have to consider the facilities offered for regular modification of characters, for the purpose of expressing common combinations. The best device for this purpose is the small circle or loop, which may be attached to the characters in any position with the utmost facilty. Beside this there are the small hooks ⌣ and ⌢ , which may be written with almost no expenditure of time at the upper or lower ends of characters respectively. There are also the small circles ⋏ and ⋎ , as used in the Longhand letters *ℓ* and *f* ; which may also be freely used in an alternating system. All these "symbols" are of such a nature that they do not affect the level of the writing, and are, therefore, specially adapted in an alternating system to the purpose of expressing combined sounds, not separated by a vowel.

We can now pass to the last part of our subject, as shown in Fig. IX., the final assignment of the various signs to the different sounds. And here we must first notice two important rules, which should, as far as possible, guide our choice. The first is : " Like sounds must be provided with like signs ;" the second : " Commoner sounds must be provided with easier signs." It is obvious that these are, to a certain extent, antagonistic, and we are, therefore, forced to some kind of a compromise, and, the type of system once selected, it is upon this compromise that the excellence of a shorthand method depends. The principle that seems to be the best to follow in making this compromise is to observe most closely the first rule, because upon this depends the possibility of abbreviating by regular and comprehensive rules. The second rule will decide what *class* of signs we give to what *class* of sounds. For this purpose it is important to know what sounds, and what classes of sounds, are the commoner. We have often been told what *letters* are most frequent, but the question of the commonest *sounds* has been rather neglected. After counting the sounds in a great number of words, I find that they arrange themselves as follows :—

I.—Vowels.

   (a) Simple Vowels :—

ă	17	times per 100 words.		
ĕ	15	,,	,,	,,
ĭ	33	,,	,,	,,
ŏ	16	,,	,,	,,
ŭ	23	,,	,,	,,
oo	6	,,	,,	,,

(*b*) Compound and "Long" Vowels.

aw	3 times per 100 words.			
ay	5	,,	,,	,,
ē	6	,,	,,	,,
ah	3	,,	,,	,,
ur	1	,,	,,	,,
oo	4	,,	,,	,,

In the above calculation the number of instances of the vowel ŭ includes the cases where the termination ŭ follows a long vowel, as in *our*, *your*, *air*, &c. It has been decided above to deal with these cases in a special manner, and if we exclude these, ŭ will be found to come after ă and ĕ in frequency. The three commonest vowels, then, are ă, ĕ, and ĭ; the three next, ŏ, ŭ, and oo. The compounds ō, oi, ī, and ow, are very much less common. On comparing the table on page 25 with that on page 26, and considering the comparative frequency of the sounds, we are at once struck with the fact that the straight liaison should be devoted to ă, ĕ, ĭ, and their "long" vowels; the curved liaison to ŏ, ŭ, and oo; and the double-curved liaison to the vowels ō, oi, ī, and ow.

II.—CONSONANTS.—The commonest sounds are n (37 times per 100 words), t and d (20 times per 50 words), r (15 times per 100), s and z (13 times per 50), and l (10 times per 100). These are all consonants which occur constantly in combination, and if the cases of combined consonants, which are to be dealt with by special rules, be eliminated, it will be found that m, f and v, ng, p and b, k and g, are nearly as common as those already mentioned. If the table on page 25 is compared with that on pages 27 and 28, it will be observed that a good deal of alteration is necessary before they can fit into each other, but that, roughly speaking, and taking into consideration the demands of the second rule, the first four lines in the second diagram are well adapted to represent the four lines of the first. It would take too much space to enter fully into the compromise which is necessary, but the result, as shown in Linear Shorthand, gives us the following convenient arrangement :—

P, B, //	F, V, ‹ ‹	M, W, 〉 〉	R, L, ‹ ‹
T, D, //	Th, Dh, ‹‹	N, Ng, 〉 〉	S, Z, ‹ ‹
K, G, ‹ ╲	Ch, J, ╲ ╲	H, Y, ╱ ╱	Sh, Shon ╱╱
Pr, Br, ‹ ‹	Fr, Vr, ╲ ╲	Pl, Bl,))	Kr, Gr, ╲ ╲
Tr, Dr, ‹ ‹	Thr, Dhr, ╲╲	Fl, Vl,))	Kl, Gl, ╲ ╲

There remain for consideration the modifiers, or symbols representing coalescing consonants and vowels. Some of the coalescing sounds occur, as has been shown on page 25, most frequently in certain positions with regard to the character with which they combine, and with regard to the word. In the same way, as we saw on page 28, some of the " symbols " are most easily formed in certain positions. This leads us inevitably to the following arrangement of the modifiers :—

s, occurring in all positions, is represented by the symbol   o   .

n, occurring generally as the first of a pair, by the symbol   ⌣   .

t, occurring generally as the second of a pair, most often at the end of
        words, by the symbol   ⌢   .

w and y, occurring generally as the second of a pair in the middle of
        words, by the symbols   ⤲   ,   ⤳   .

The vowel termination ŭ, occurring as a modifier at the end of vowels,
        by the symbol   ◂

In conclusion, I would repeat that these three papers on the Science of Shorthand were prepared at a time when I was using a system of the Cursive type, which had proved most successful in note-taking, and in other ways. It was naturally with reluctance that I was forced to the conclusion that the soundest basis for a Shorthand system was the alternating principle. The arguments here set forth are not, therefore, special pleading on behalf of Linear Shorthand, but a literal account of the reasons for its inception, and of the manner in which it was designed.

A MANUAL OF LINEAR SHORTHAND.

PART I.—CORRESPONDING STYLE.

TABLE OF CONTENTS.

	PAGE
PREFACE.—Origin and Character of Linear Shorthand	i
INTRODUCTION.—The Science of Shorthand.	
Part I.—The Sounds of Speech	1
Part II.—The Signs of Writing	9
Part III.—Combination of Sound and Sign ...	21

CHAPTER I.—Consonants (Downstrokes).

Lesson I.—Table of Consonants	33
Lesson II.—Joining the Consonants	34

CHAPTER II.—Vowels (Upstrokes).

Lesson III.—ă, ŏ, Ĭ ; aw, ay, ee	35
Lesson IV.—ŏ, ŭ, oo ; ăh, ur, oo	39
Lesson V.—Word signs and Arbitraries	41

CHAPTER III.—Vowels—continued.

Lesson VI.—i, ow, oi, o	45
Lesson VII.—ur-Vowels	47
Lesson VIII.—Initial Vowels	49
Lesson IX.—Final Vowels	50

CHAPTER IV.—Consonant Grouping.

Lesson X. —Double Consonants	55
Lesson XI.—Coalescents (Horizontal Characters)	56
Lesson XII. - Coalescents—continued	60
Lesson XIII.—Accidental Consonant Combinations	67
Lesson XIV.—Neutralization of Vowels	70

CHAPTER V.—Abbreviation.

Lesson XV.—Word Signs	73
Lesson XVI.—Prefixes	78
Lesson XVII.—Suffixes	83
Lesson XVIII.—Phrasing	86

CHAPTER VI.—Examples.

CHAPTER I.

LESSON I.

TABLE OF CONSONANTS.

Two sizes of characters are distinguished in Linear Shorthand, as in Longhand. In the downstrokes two thicknesses are also employed to distinguish between nearly identical sounds. The following Table is to be copied again and again, until it can be written easily from memory, and the signs quickly and neatly formed. A fine flexible pen, preferably of gold, and good ruled paper should be used, and the consonants sounded with a faint neutral vowel sound following them.

Exercise I.—Copy the following Table :—

P.B.	F.V.	M.W.	R.L.
T.D.	Th.Dh.	N.Ng.	S.Z.
K.G.		H.Y.	
	Ch.J.		Sh.Shon.

Exercise II.—(A.) Write in Longhand :—

(*B.*) Write in Linear Shorthand:—

Sh, S, L, J, V, Dh, N, M, Y, D, G, R, P, H, Ch, Shon. F, V, T, Z, Ng,
K, Th, W, S, P, T, K, B, D, G, F, Th, Ch, M, Dh, B, W, H, Y, R, Ng,
Shon, N, Sh, L, J, Z, Sh, P, W, M, S, H, Dh, L, Ch, P, J, Shon, T, W, Z,
V, F, K, H, Y, B, Dh, N, T, D, M, Z, G, R, Ng, F, Y, D, K, Th, Shon,
N, Ch, Th, R, W, Sh, Dh, S, Z, Sh, L, B, K, H, W, Y, J, V, Shon, Ng.

(*Note:—The second part of every Exercise is a key of the first part.*)

LESSON II.

Joining the Consonants.

The signs of Lineal Shorthand, as its name implies, are placed side by
side on the line, as in Longhand, and joined by an upward connecting
stroke, called the *liaison*. In joining the characters $\int\int \int\int\int$, a loop is
introduced thus, p-sh $\mathscr{S}$: y-l $\mathscr{N}$; h-shon $\mathscr{S}$.

Exercise III. —(*A.*) Write in Longhand:—

(*B.*) Write in Linear Shorthand :—

p-w, h-p, t-v, r-l, k-b, f-t, th-w, ch-f. m-v. n-m. s-r, su-p, b-z. y-dh,
d-n, l-ng, y-z, v-th, dh-r, j-l, ng-w-dh. l-ng-g, b-z-h, j-sh-ng. g-w-g,
r-p-d, z-h-g, b-t-r, m-r-y, h-k-ch, th-f-shon. p-th-s, w-ch-n, s-m-ch.
d-n-b, k-s-d, y-sh-z, m-w-y, r-y-ch.

CHAPTER II.

LESSON III.

ă, ŏ, ĭ; a͞w, a͞y, e͞e.

Before proceeding to the consideration of the vowels it is necessary to say something as to the spelling adopted in Linear Shorthand. Words are represented according to their *sound*, not according to the letters by which they are spelled in ordinary writing. Thus, *right*, *write*, and *rite*, are all written rīt, while *tear* (of crying), and *tear* (to rend), *bow* (a weapon), and *bow* (an inclination), are differentiated as in speaking. As regards legibility there can be no question of the superiority of the Phonetic spelling over the ordinary orthography, inasmuch as the former actually represents words as they are heard in conversation, while the latter only records (very imperfectly) their philological history, and makes it necessary that the mode of representing each word, or class of words, should be learnt almost as if it were an arbitrary sign. For Shorthand purposes, the importance of having a phonetic alphabet has been sufficiently shown in the Introduction. It depends, first, upon the necessity of orderly arrangement, and secondly, upon the necessity of representing words in the most economical way. This method of spelling has accordingly been adopted in nearly all systems since Pitman's Phonography. Its great disadvantage is that many persons find a difficulty in divesting their minds of all thoughts of the ordinary spelling, which has been acquired with so much trouble in childhood. They fail, for instance, to notice that the r in *warm*, *sport*, *here*; the l in *half*, *would*, *walk*; the g in *right*, *caught*, *dough*; the a in *foal*, *teak*, *beauty*; the i in *daisy*, *fruit*, *piece*, &c., &c., are not present in the sound of the word. It is a good plan to begin by sounding aloud every word as it is written, and to think, if necessary, of other words with which it rhymes. The difficulty is one which, though often serious at the outset, disappears very rapidly as the student advances, and learners are therefore warned not to become discouraged if at first the method of representing words in Linear Shorthand seems puzzling. Having said so much, we can now proceed to the representation of the vowels.

It was said in the last lesson that the consonants were connected by an upward connecting stroke, called the "*liaison*." In Longhand the *liaison* is without meaning, in Linear Shorthand it indicates the various vowels.

ĕ. When two characters are written close together and the second stands on the line, the vowel ĕ, as in **pet**, is indicated between them. Thus :—

[shorthand] pet.   *[shorthand]* ten.

[shorthand] fed.   *[shorthand]* sell.

[shorthand] hedge.   *[shorthand]* wreck.

This same connection is used for the neutral vowel in such words as *mettle, rebel,* &c. Thus :—

[shorthand] mettle (*or* metal).   *[shorthand]* rebel (noun).

[shorthand] wrestle.   *[shorthand]* revel.

ay. When the characters are written further apart, with the same type of connection, the "long" form of the vowel, viz.: ay, is indicated between them. Thus :—

[shorthand] pate.   *[shorthand]* bathe.

[shorthand] sale.   *[shorthand]* rake.

[shorthand] remain.   *[shorthand]* cable.

Exercise IV.—(*A.*) Write in Longhand :—

[shorthand lines]

(*B.*) Write in Linear Shorthand :—

> pet, set, fell, Ned, Bess,
> when, yes, then, shed, Jess,
> mace, pate, sate, fail, paid,
> base, wane (*or* wain), face, bane, jays,
> shade, relation, chain, bathe, save,
> wage, guess, yell, days, said,
> settle, cable, meddle, retain, revel,
> taken, faces, terrace, Bennett, Kenneth,
> relate, wretched, vessel, belated, renegade,
> delayed, cessation, venerate, senate, relegated.

ă. When two characters are written close together, and the second is slightly *above the line*, the vowel ă, as in **pat**, is indicated between them

Thus :—

_____ pat..   _____ fan
_____ fad.   _____ Sall
_____ hatchet.   _____ cattle

This same connection is used for the *short sound* of **aw**, which occurs in English only before l followed by a surd, *e.g.*, *salt, false*. In such a position the vowel must be read **aw** instead of **a**. This combination cannot be illustrated till Lesson XI. (See Note, page 58.)

aw When the characters are written further apart, with the same type of connection, the long vowel **aw**, or **or**, is indicated between them. Thus —

_____ taught.   _____ call.
_____ Saul.   _____ pause.
_____ borne.   _____ mourn.
_____ wart.   _____ thought.

Exercise V.—(*A.*) Write in Longhand :—

(B.) Write in Linear Shorthand :—

bang, fan, cad, ham, than,
hath, rack, sham, mash, ran,
bought, port, fought, caught, hoard,
chawed, walk, shawn, wars, fawn,
sword, shawl, chalk, gnawed, ball,
rattle, pallet, damage, rattan, Cavan,
savages, fallen, channel, Rahab, pauses,
caution, caravan, palaces, reported, generation,
catamaran, carat, portal, recorded, shorten,
parade, garret, canal, bedad, Jordan.

1. When two characters are written close together, and the second is slightly *below the line*, the vowel **I** as in **pit**, is indicated between them. Thus :—

_____ pit.   _____ tin.
_____ letting.   _____ taking.
_____ radish.   _____ Phyllis.

449524

ĕ. When the characters are written further apart, with the same type of connection, the *long* form of the vowel, viz.: ē, is indicated between them. Thus :—

    peat (or Pete).   *been (or bean)*.

    lateen.   *detail (noun)*.

    feeble.   *leading*.

When the syllable ing follows a long vowel, the l is omitted. Thus :—

    saying, (*lit :* sāyng).

    gnawing, (*lit :* nāwng).

    being, (*lit :* bēēng).

Exercise VI.—(*A.*) Write in Longhand :—

[shorthand characters]

(*B.*) Write in Linear Shorthand :—

> pith, fit, thin, midge, rig,
> chin, give, this, fish, hill,
> beat, teak, reach, seem (*or* seam), deal,
> reef, sheet, heal (*or* heel), bees, zeal,
> with, these, being, peel, pawing,
> people, village, David, believe, pouring,
> radishes, vegetate, beetle, chalice, valleys,
> fording, native, receiving, details, needed,
> validate, division, reveal, rallied, wishes,
> recording, belaying, civil, yelling, redeem.
> This will perish ; that shall remain.
> Take heed that these men receive him well.
> Jim will wait till his relative marries.
> When did these calamities happen ?

LESSON IV.

ŭ, û, o͝o : âh, u͡r, o͡o.

The six vowels learnt in the preceding lesson have been indicated by help of a *straight liaison*; an equal number, in similar positions, are indicated by help of a *curved liaison*. The following rules apply to all positions and distances alike :

(i.) After a consonant ending without a hook, as / ／ . the curved liaison is commenced straight upward. in the line of the consonant, thus :—

/ , ／ , ૪ .

(ii.) After a consonant ending with a hook, as ɩ ℓ , the curved liaison is commenced by widening the hook, thus :— / ′, ? . ＼ ,

(iii.) Before a consonant beginning without a hook, the latter part of the curved liaison ascends the line of the consonant, thus :— ⁄, ⁄ℓ , ⅃.

(iv.) Before a consonant beginning with a hook, the latter part of the curved liaison is formed by widening the hook. thus :— /⁄, ℓ, ＼.

The vowels indicated by the curved liaison are as follows :—

ŭ. Second character *on the line*—close connection.
ur. ,, ,, ,, wide connection.

Thus :—

but.	sum.
.rough.	shun.
—mirth.	learn.

ŏ. Second character *above the line*—close connection.
âh. ,, ,, ,,. wide connection.

Thus :

.pod.	forage.
rot.	poll (parrot).
farm.	part.

oo. Second character *below the line*—close connection.
ōō. ,, ,, ,, wide connection

Thus :—

.put.	puss.
woman.	foot.
look.	food.

oo is often read yeo, especially at beginning of words.

Exercise VII.—Copy out the following Table till it is thoroughly familiar.

		CLOSE CONNECTION.		WIDE CONNECTION.	
STRAIGHT LIAISON.	Above line On line Below line	ă 	 ĕ I	āw 	 āy ē̄
CURVED LIAISON.	Above line On line Below line	ŏ 	 ŭ oŏ	âh 	 ūr oō

Exercise VIII.—(*A.*) Write in Longhand :—

(shorthand characters)

(*B.*) Write in Linear Shorthand :—

      pun, pup, pen, dead, touch,
      wretch, some (*or* sum), leg, huff, hung,
      turn, pain, bird, curl, firth,
      faith, nurse, worm, main, work,
      doff moth, Don, Dan, fog,
      dog, bag, hog, shock, botch,
      tarn, torn, hearth, baul, hark,

parch, porch, lard, charm, birth,
put, pit, look, luck, lick,
good, bush, tush, rash, cook,
doing, fume, suit, seat, whose (*or* hues, hews),
shoes, cool, boot, foot, food,
knowledge, charters, vermifuge, residues,
religion, resolute, vicious, hurries,
foreign, tattooing, collusion, charming,
certain, territories, fortitude, venomous,
porridge, martin, covet, subacid,
choose (*or* chews), juice, Jews, circulating,
Burberris, Volapuk, Hercules, Shibboleth.

LESSON V.

WORD-SIGNS AND ARBITRARIES.

Most of the consonant forms, when standing alone, are used to represent short common words which contain them. The following rules govern the selection of these words.

(i.) A consonant standing alone on the line represents a word beginning with that consonant, and followed by a vowel.

(ii.) A consonant standing alone, above, or below the line, represents a word beginning with ă or ĭ respectively (in some few cases preceded by **h** or **w**), followed by that consonant.

The *word-signs* to be learnt at present are as follows :—

the figure one, as in Longhand,

╱ be (*rarely* been), ·ι for, ℓ very, ╱ me (my, *or* may), ╱ we (*or* were), ι let,

╱ to, ╱ do, ℓ thing (*or* think), ℓ there (*or* they), ╱ no, ℓ so,

ι can, ↖ go, ∫ who, ∫ ye (your, *or* year),

↖ gentleman (*or* -men), ∫ shall, *or she*.

ʿ have,   *?* am,   *ɩ* all,

.ʃ at,   *.ʃ* had,   *ʃ* hath,   *ʃ* as (*or* has),

ι if,   *ɔ* him.   *ɔ* ill (*or* will),

/ it,   */* in,   */.* England (*or* English),   */* with,   */* is (*or* his),

\. which.

Exercise IX.—Copy out the above table, till the word-signs can be written out, in their order, from memory.

The following *arbitrary signs* should next be learnt :—

        -        a, an, and.

        ˎ   the.

        -        of.

They are **written as small as possible**, and may be joined to one another, thus :—

        ⌒   and of the.

        ˏ   and the.

        ⌐   of a.

A dot above the line placed close to the end of a word adds the termination-ly, as :—

    2ʃ· ____sadly,   *14·* ____tacitly.

    1⁄2ᴄ ____naturally.   *14·* ____vapidly.

Proper names are represented as a rule in Longhand; if written in Shorthand a tick is struck diagonally under the first letter, thus :—

    ɭ₂ ____Tom.   *ↄʃ* ____Phyllis.

    ιʃₑ ____Riddell.   *ɛɭ* ____Harden.

Punctuation and *numerals* are written as in Longhand.

Exercise X.—(*A.*) Write in Longhand :—

[shorthand characters]

(*B*.) Write in Linear Shorthand :—

> visionaries, very, capital, with, all,
> thoroughness, rebel, where, rudely, ye,
> participation, can, revenues, his, caparison,
> been, pulleys, if, literally, in,
> Devenish, who, jetison, ill, covering,
> were, Puritan, England, caravan, go,
> barrack, year, Paladin, am, Nemesis,
> to, Caliban, shall, which, no,
> villain, think, salaried, me, cinnamon,
> your, Tunis, so, noodle, thing.

(*C*.) Read aloud :—

[shorthand characters]

(*D.*) Write in Linear Shorthand :—

ALFRED AND THE CAKE.

In the year 877, there was living in the marshes in a savage part of England, a woman, who has won a name in the record of this nation, as having made a famous cake. It was the sort of rude cake which was the common food of the rural population at that date, and when she had made it she put it at the hearth to bake, calmly telling the serving man to sit and watch it, so that it should not burn. This serving man was, indeed, the King of England, and had very much to think of, as the Danish forces were searching for him to put him to death. So the King in his peril forgot to watch the cake as the woman had bidden him, and it did burn. When the good woman came back, she gave a look at the King, and the cake burning at the hearth, and was of course very wroth with the serving knave, as she thought him, and began to call him all the hard things (*write* "thing") she could think of ; calling him a good for nothing varlet, for having let the cake take harm. She was jeering at him, and mocking him, for sitting, as she said, "as if it were a gentleman, forsooth, who had all the calamities of a nation to support," when suddenly they heard a horn at the gate of the yard, and all the leaders of the English came in, and, kneeling to the King, began to tell him that the Dane had been defeated in battle some days back, and was then suing for peace.

[*The above passage contains* 270 *words. The learner should not pass on till he can write it neatly from dictation in ten minutes.*]

CHAPTER III.

VOWELS – (CONTINUED).

LESSON VI.

i, ow, oi, o.

There remain four vowels for which we have not as yet provided a means of expression. These are the vowels ī, ōw, ōī, and ō. as in buy. bough, buoy, bow (archer's). These are indicated by help of a *double-curve liaison*, *i.e.*, a liaison whose curvature is broken in the middle. As a rule, the break in the curvature is turned upwards. thus / , : the only exception is where the preceding consonant ends without a hook and the following consonant commences with a hook, thus / .

The vowels indicated by the double-curve liaison are as follows :—

ī.   Second character *on the line*—close connection.
ōw.   „   „   „   wide connection.

Thus :—

	bite.		life.
mile.		guide.
mouth.		town.

ōī.   Second character *above the line*—close connection.
ō.   „   „   „   wide connection.

Thus :—

	buoyed.		foil.
Doyle.		Yoick.
coal.		home.

No use can be made of the double-curve liaison with the second character *below the line.*

Exercise XI.—(*A.*) Write in Longhand :—

(*B.*) Write in Linear Shorthand :—

> Light, shine, joys, knows. shout,
> kine, foil, pouch. touch, those,
> gyrate, toiling, devoted, selenite, thought,
> powderous, buying, votaries, localization,
> Simon, repose, polling, doubters, joying,
> refusal. surmises, mobilize, Maynooth,
> meadows, Visigoth. boycotted, rowing, hiding,
> widowed, turmoil, chosen, final, definitely,
> title. Norfolk. folk, loitering, pilot,
> noisome, synagogue, rogue, zymotic, notices.

(*C.*) Read aloud :—

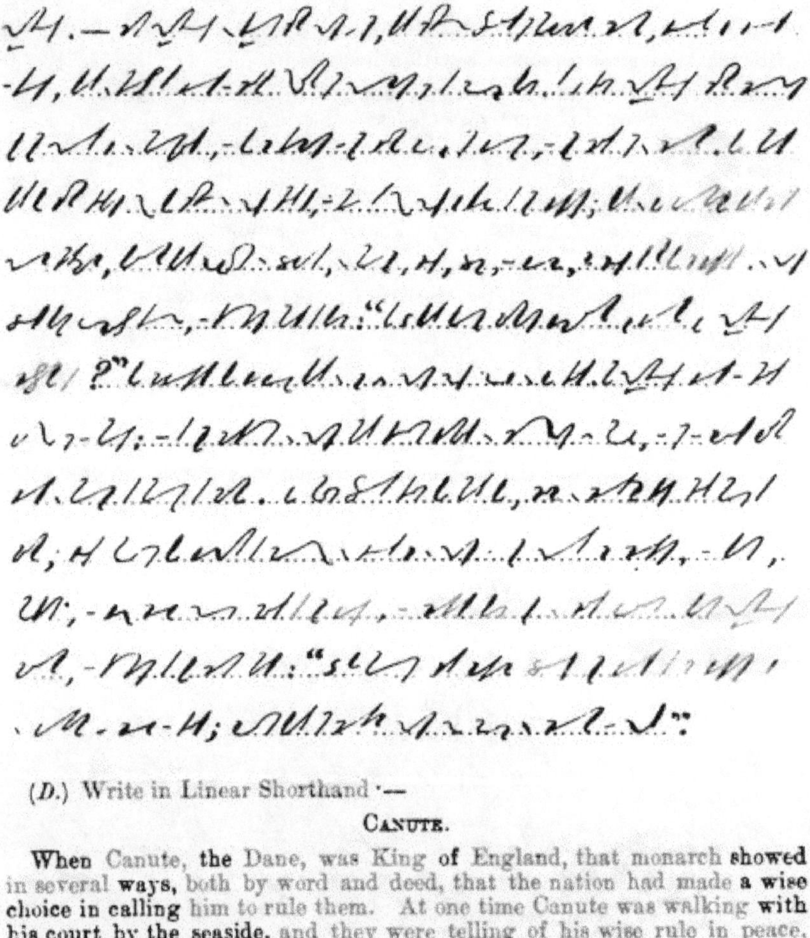

[shorthand text]

(*D.*) Write in Linear Shorthand ·—

CANUTE.

When Canute, the Dane, was King of England, that monarch showed
in several ways, both by word and deed, that the nation had made a wise
choice in calling him to rule them. At one time Canute was walking with
his court by the seaside, and they were telling of his wise rule in peace,
and his might in the wars. They said that there was nought which their

monarch could not do, and nothing that could be deaf to his bidding : that
the very forces that men called natural, those that govern the heaven,
the seas, wood, hill, and vale, all would do as they were bidden. The king
heard this foolish talk, and turning, said to them : "Think ye that these
rising billows will pause if Canute wishes it?" They repeated their
belief that the will of the king could curb the very tide. So Canute bade
a knight fetch him a seat ; and at his return the king sat down beside the
margin of the surf, and, in a loud voice, bade the seas to cease to rise.
For some short time they sat there, while the waters did not seem to rise ;
but soon they began to reach the part where the king and his court were
waiting, and then, suddenly, a big wave came right to his feet, and wetted
them with the white foam. Then Canute rose, and, turning to his men,
said : "Ye have seen what little heed is paid to my bidding by the forces
of wave and tide ; learn that no mortal king can rule the ways of God."

[*The above passage contains* 270 *words ; the learner should not pass on till he
can write it neatly from dictation in ten minutes.*]

LESSON VII.

" *Ur* "-Vowels.

There is a peculiar class of compound vowels, which is formed by the
addition of the termination ŭ to the vowels already considered, both short
and long. Instances of these " *ur*-vowels " are those in bare (= ă ŭ),
fear (= ī ŭ), dire (= ī ŭ), grower (= ō ŭ). This termination is indicated
by a very small *left-handed* loop at the end of the liaison which indicates
the vowel, and above and to the left of the following consonant. It is
similar to the small loop at the beginning of the longhand letter ↓ .

We thus write :—

pierce. Sayers.

fared. leeward.

The following rules must be observed in writing the ur vowels :—

(i.) A *curved liaison*, when followed by the termination ŭ, is written as if
the succeeding consonant did not commence with a hook, whether it does
so or not. This is obviously necessary, to enable the ŭ loop to be made on
the right side.

Compare :—

bourne, *with* boon.

lures, *with* loose.

[*NOTE.*—oo *is the only vowel indicated by single-curve liaison which is ever
followed by the termination* ŭ.]

(ii.) When a *double-curve liaison* is followed by the ŭ termination, the latter is indicated by introducing a small loop or circle at the break in the curvature, instead of at the upper end of the following consonant.

Thus :—

	buyers, *not*	
	fired, *not*	
	tired, *not*	
	coward, *not*	

(iii.) The "*ur*-vowels" are occasionally employed to represent combinations of the primary vowels with other terminations than ŭ. Thus :—

Hewitt.   poet.

(iv.) When a long vowel is followed by *trilled* r, the ŭ termination is nearly always introduced in speaking, but need not be written. Thus :—

weariness.   furies.

lurid.   mayoress

Exercise XII.—(*A.*) Write in Longhand :—

(*B.*) Write in Linear Shorthand :—

laird, Bayard, real, boors, Bowen,
cowardly, fired, poet, towers, sewers,
weird, wired, layers, dared, pierce,
mares, leeward, fires, bowered, rowers,
lowered, Howard, showered, retiring, bayonet,
fears, cairn, bearing, riot, royal,
nearing, bourne, Coan, tiresome, violate,
theoretical, mayoress, Puritan, dialogue, fairies,
fearlessly, powerful, repaired, coalition, curate,
devoured, Naaman, pieties, Caribbean, dowries.

LESSON VIII.

INITIAL VOWELS.

(i.) INITIAL VOWELS are indicated in the same way as medial vowels, the liaison being commenced on the line, thus :—

_____ pet.   _____ et.

_____ pate.   _____ eight.

(ii.) The liaison is always omitted before ă and ĭ, as being unnecessary, thus :

_____ pat.   _____ at.

_____ pit.   _____ it.

but :—

_____ peat.   _____ eat.

(iii.) Initial curved liaisons are written in their simplest form, that is, before a consonant beginning without a hook, the left-handed curve _____ , is used ; before a consonant beginning with a hook, the right-handed curve _____ , is used, thus :—

_____ office.   _____ honours.

_____ uttered.   _____ earning.

_____ youth.   _____ eulogize.

(iv.) Initial double-curve liaisons are not used. The vowels indicated by double-curve liaisons medially are indicated initially (and finally) by the curves _____ and _____ ; the former being used before consonants beginning without a hook, the latter before consonants beginning with a hook. Thus :—

_____ item,   *not* _____

_____ isolate,   *not* _____

_____ "outing,"   *not* _____

_____ owlet,   *not* _____

_____ oaten,   *not* _____

_____ oiliness,   *not* _____

_____ ozone,   *not* _____

This rule will not be found to cause any ambiguity, but it is as well to point out that—

_____ = Arth·   _____ = bath.

                   _____ = oath.   _____ = both.

_____ = us.   _____ = fuss.

                   _____ = ice.   _____ = vice.

The general principle governing all these cases is that the *simplest distinctive curve* is used.

(v.) Initial *wr*-vowels are very rare; they are indicated in the same way as when they occur medially, thus :—

       "airth."   ewers.

       iron.   hours.

Exercise XIII.—(*A.*) Write in Longhand :—

(shorthand)

(*B.*) **Write in Linear Shorthand :—**

> appearing, iterate, opposite, Eutychus, eighteen,
> eaten, urban, Idomeneus, Airedale, enemies,
> inimical, honouring, uneducated, emus, ermine,
> armories, united, oval, ominous, aerated,
> Europe, ironing, ocean, youth, earthen,
> Arthur's, acres, awful, oven, effusively,
> avaricious, esoteric, ushered, organize, **azure**,
> oozing, ironical, Yucatan, irk, Argolis,
> eagerness, orchard, occupied, irrigate, **acolyte**,
> ashes, Eumenides, islet, wiliness, officers.

LESSON IX.

FINAL VOWELS.

(i.) Final vowels are indicated in the same way as medial vowels, the liaison being written as if followed by / or / in the required vowel position, thus :—

     pate.   pay.

     feet.   fee.

     port.   paw.

(ii.) Final curved liaisons are written in their simplest forms, that is, after a consonant ending without a hook the right-handed curve, ⌒ , is used; after a consonant ending with a hook the left-handed curve, ⌒ , is used, thus:—

　　　　　　　　　bar.　　　　　　　far.
　　　　　　　　　purr.　　　　　　　fur.
　　　　　　　　　mew.　　　　　　　rue.

(iii.) Final double-curve liaisons are not used. The vowels indicated by double-curve liaisons medially, are indicated finally (as initially) by the curves ⌒ and ⌒ ; the former being used after consonants ending without a hook, the latter after consonants ending with a hook, thus:—

　　　　　　　　　deny, *not*
　　　　　　　　　terrify, *not*
　　　　　　　　　how, *not*
　　　　　　　　　cow, *not*
　　　　　　　　　toy, *not*
　　　　　　　　　joy, *not*
　　　　　　　　　woe, *not*
　　　　　　　　　fellow, *not*

(iv.) Final *ur*-vowels are indicated in the same way as when they occur medially, thus:—

　　　　　　　　　dare.　　　　　　　fear.
　　　　　　　　　boor.　　　　　　　fewer.
　　　　　　　　　tower.　　　　　　fire.

Compare:—

　　　　　　　　　dared.　　　　　　feared.
　　　　　　　　　bourne.　　　　　　lured.
　　　　　　　　　towers.　　　　　　fired.

(v.) Vowels standing alone follow, of course, the rules of both initial and final vowels, thus:—

　　　　　　　　　or.　　　　　　　eh.
　　　　　　　　　are.　　　　　　　you.
　　　　　　　　　I.　　　　　　　(h)ow.
　　　　　　　　　our.　　　　　　ire.

Curved liaisons when standing alone begin with a right-handed curve, as shown above; so ⌒ , *not* ⌒ ; ⌒ , *not* ⌒ ; ⌒ *not* ⌒ .

(vi.) When two vowels occur together in a word (a coincidence very rare indeed), they are written disjoined, but close together, the first being written as if final, the second as if initial, thus:—

　　　　　　　　　oasis.　　　　　　Bœotia.

Exercise XIV.—(*A.*) Write in Longhand :—

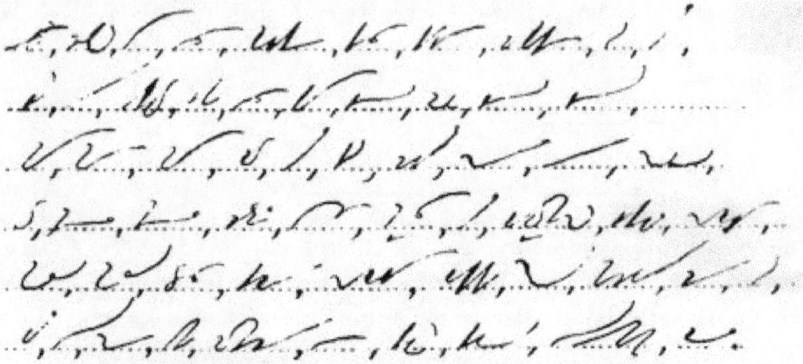

[shorthand content]

(*B.*) Write in Linear Shorthand :—

Asia, mother, are, our, Zebedee,
dower, deter, residue, I, air,
boy, or, edify, apathy, hour,
though, bough, rough, bow (to incline), bow (of an archer),
sow (to plant), sow (a pig), sew, alloy, air.
dye, repair, gayer, eh, curfew,
she, inure, newer, warily, armour,
Noah, heir, Veronica, Belvidere, calibre,
sour, sower, shower, delay, Calabar,
avidity, jar, semaphore, weigh, eye,
affair, oh, quay, any, runaway,
you, differ, defer, oasis, allure.

(*C.*) Read aloud :—

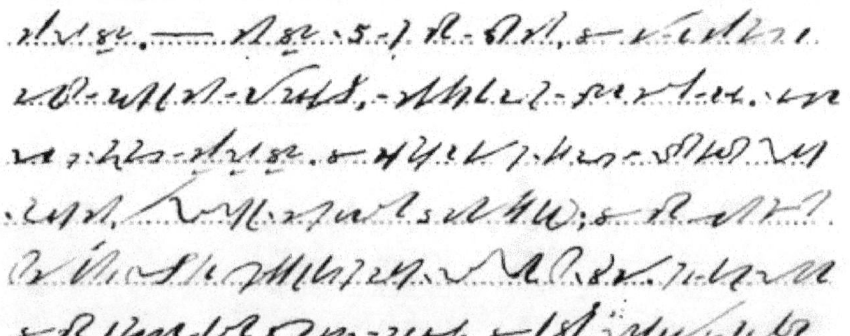

[shorthand content]

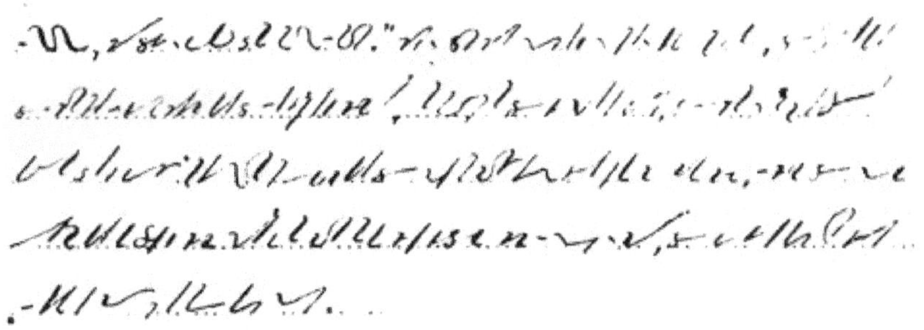

[shorthand writing]

(*D.*) Write in Linear Shorthand :—

MAD-CAP HARRY.

When Harry V. of England was a young man he bore a very bad name by reason of living with men of low reputation, and imitating their loose and immoral mode of life. The people gave him the nick-name of Mad-cap Harry. He would sit all day in the tap-room of a common tavern chaffing the serving men, or joking with the mean fellows who resorted thither; he was even known on more than one occasion to have united with them in robbing the coaches on the high-way. In one of these cases, he was, with several of those whose peril, and maybe booty, he had shared, carried before one of his father's judges. During the course of the case, Mad-cap Harry is said to have hit the judge in a sudden fit of fury, whereupon the latter committed him to gaol as if he were any ordinary felon. The young man, knowing that he had been in the wrong, permitted the officers to lead him away without any opposition. When the King heard of it, he said, "Happy is the monarch who has such a judge, more happy the father who has such a son."

When the young man came to be King at the death of his father, he showed that he was not really of the base metal that he had appeared to be while heir. As soon as he began to reign, he was careful to show those who had been formerly his teachers in evil that he could no longer take part in their lawless life, and while he gave orders that they should be well cared for as long as they would behave well and keep the law, he forbade them on pain of death to allow him to see them again.

[*The above passage contains* 304 *words; it should be written in ten minutes.*]

CHAPTER IV.

Consonant Grouping.

No means have as yet been provided for uniting two consonants coming together without an intervening vowel. This chapter will be devoted to this purpose, and to the means by which the grouping of consonants may be shown.

The manner in which the consonants are disposed in words, some of them grouped together, others standing singly, separated from each other by vowels, is an important feature of the English language, so much so indeed that when clearly shown it enables us to dispense very largely with exact differentiation of the vowels. It is therefore most necessary that a shorthand system should provide such means for representing this grouping of the consonants as may be used without hindrance at the highest speed. The fundamental principle of Linear Shorthand is that consonants are represented by downstrokes, and vowels by upstrokes; hence it is obvious that in a consonant combination no upstroke must be seen, and that, as a general rule, no combination should occupy more space, vertically, than the single consonant signs. Therefore, except in the case of rare accidental combinations, it must be arranged that either the combination is represented by a simple downstroke of the same height as a consonant, or else the secondary part of the combination is indicated by a horizontal symbol. Now if we examine a list of consonant combinations we shall find that they fall naturally into three classes. First we have a series of strong double consonant sounds, combinations of r and l with the others, of very common occurrence, and found chiefly at the commencement of words. Each of these common pairs is, in Linear Shorthand, provided with a single simple sign, corresponding in size with that of the principal sound of the combination. Next we have a number of groups formed by the combination of certain six consonants, s, n, t, d, w, and y, with the others, either as the first or second part of the combination. These *coalescent* consonants are provided with *coalescent* symbols, small loops and hooks, which combine with the sign to which they are attached, and do not affect the lineality of the system. Lastly, we find that occasionally other consonants occur in juxtaposition, brought together accidentally, generally at the point of junction in compound words. These are represented by their alphabetical characters joined together without an upstroke.

The author desires emphatically to point out that the special characters used for the representation of Consonant Groups in this system are not "alternatives," such as are used in other systems. In Linear Shorthand no word can be fully written in more than one way.

LESSON X.

Double Consonants. (First Class of Consonant Combinations.)

The double consonants are represented, like the single consonants, by simple down strokes, arranged in sets of four, differentiated by size and thickness. The following table is to be copied again and again until it can be written easily from memory, and the signs quickly and neatly formed. The consonants must be sounded with a faint neutral vowel sound following them, thus, prŭ, trŭ, &c.

Exercise XV. Copy the following Table :—

Pr. Br. ϲ ϲ	Fr. Vr. ⌣ ⌣	Pl. Bl. ﹚ ﹚	Kr. Gr. ﹨ ﹨
Tr. Dr. ϲ ϲ	Thr. Dhr. ⌣ ⌣	Fl. Vl. ﹚ ﹚	Kl. Gl. ﹨ ﹨

[It may be noted that the tabulation of double-consonants for representation is not perfectly regular, since **Fl** and **Kl** take respectively the places theoretically belonging to **Tl** and **Chr**, which are not required.]

These characters are joined to other consonants by liaisons in the same way as the single consonant characters. The characters for **kr**, **gr**, and **kl**, **gl**, begin and end without hooks, like those for **p, b, t,** and **d,** and are joined in the same way. No loop is introduced in joining them as in the case of the characters for **h, y, sh,** and **zh.** Thus :—

decrease.   reclaim.
reckless.   Raglan.

The characters for **fr, vr, thr, dhr,** begin without a hook, and end with a hook, like those for **f, v, th** and **dh,** and are joined in the same way, thus :—

taffrail.   through.
Ivry.   saffron.
throat.   sev'ral.

The characters for **pl, bl, fl, vl,** commence with a hook, and end without a hook, like those for **m, w, n, ng,** and are joined in the same way, thus :—

enabling.   blies.
flower (or flour).   poplar.
pleasing.   flash.

The characters for **pr, br, tr, dr,** are the only characters in the alphabet which commence with a backward hook, like the Longhand letter

They nearly always occur at the beginning of words; when medial they are written disjoined, like *C* in Longhand, thus:—

_ _*cv̀z*_ _ prevail.　　_ _*Ở*_ _ Trinity.

_ _*eꞥ*_ _ bracken.　　_ _*Ợ*_ _ dreary.

_ _*ul*_ _ repress.　　_ _*Ợ₂*_ _ patronymic.

_ _*s ꞥ*_ _ Hebrews.　　_ _*ul*_ _ redress.

Exercise XVI.—(*A.*) Write in Longhand :—

[shorthand characters]

(*B.*) Write in Linear Shorthand :—

> briar, freedom, pleasure, Ivry, plethora,
> filigree, trying, throwing, treble, blithe,
> leprosy, crayfish, retrograde, however, dev'lish,
> decrease, troublous, throughout, fluted, clinical,
> drawer, trifling, photograph, diagram, dockleaf,
> glories, thrice, flight, defray, privilege,
> tough, trough, through, though, thought,*
> flower, closure, gratify, trickle, sobriety,
> afraid, apron, Tapley, gravity, Drogheda,
> brethren, giggling, haply, apply, freezes.

LESSON XI.

COALESCENT CONSONANTS. (Second Class of Consonant Combinations.)

[I.] S AND Z.

We now come to the second class of consonant combinations, namely those formed by the coalescence of one of certain six consonants with the others. Of these coalescing consonants the most important is s or z (which do not need to be distinguished in the coalescent form), and we will devote this lesson to the combinations of which it forms a part.

---

* NOTE.— In the above series of five words each differs from its neighbour by but one letter, which entirely changes the whole sound.

Coalescent s or z is represented by a small circle or loop, attached to the character for the consonant with which it combines. The shape of the loop is unimportant, and is varied as is convenient. It is attached to consonants according to the following rules.

(i.) Coalescent s or z united to the hooked end of a consonant, converts the hook into a loop, thus :— ⟋ sm, ⟋ sl, ⌣ sk, ⟋ spr, ⟋ spl, ⟋ fs, ⟋ ks, as in :—

_____ small.   _____ slowly.
_____ tricks.   _____ loves.
_____ succeed.   _____ palsy.
_____ Kismet.   _____ task.

(ii.) Coalescent s or z united to the unhooked end of a consonant, forms a loop on the opposite side, thus :— ⟋ sp, ⟋ sf, ⟋ skr, ⟋ ps, ⟋ ms, ⟋ pls, as in :—

_____ sport.   _____ sphere.
_____ troubles.   _____ dubs.
_____ absence.   _____ pensive.
_____ resting.   _____ discreet.

In this case the loop naturally becomes enlarged before or after a curved liaison. Compare :—

_____ blest.   _____ rustic.
_____ fancy.   _____ answer.
_____ wisest.   _____ answered.

(iii.) When s or z occurs between, and coalescing with both of, two other consonants, the second is written straight on from where the loop ends. The following instances show the forms which are used :—

_____ = psp, as in _____ outspan.
_____ = fsp, as in _____ Shakespeare.
_____ = psm, as in _____ Apsley.
_____ = pspr, as in _____ upspring.
_____ = fspr, as in _____ offspring.

Such combinations are very rare, and may generally be divided, if convenient, into their component parts, thus :—

_____ hence-forth.
_____ back-stairs.
_____ out-stretch.

Note (i.)—The combinations s-sh, s-y, s-h, sh-s, h-s, y-s, never occur, and therefore the loop is used medially on the consonants ⟋, ⟋, ⟋, &c., without signification, to facilitate joining. At the end of words ⟋, ⟋, mean shez, shonz, thus :—

_____ wishes.
_____ nations.

Note (ii.)—It is hardly necessary perhaps to say that the circle is not used to represent s, z except when attached to consonants. It is never so used upon the vowel strokes.

Exercise XVII.—(*A.*) Write in Longhand:—

[shorthand]

(*B.*) Write in Linear Shorthand:—

> sturdy, swivel, Psyche, Jephson, vestibule,
> whence, rustics, asphodel, outspokenly, Falstaff,*
> Blavatsky, wishes, establish, ransom, Boswell,
> existence, Hesperides, divisions, diatribes, backstairs,
> dextrous, useful, precludes, risk, exert,
> falae,* falls, principle, aeronauts, Calypso,
> swathes, Shakespeare, offspring, grasping, flotsam,
> prismatic, drives, etcetera, travesty, spirited,
> acrostic, Caxton, outstretch, spherical, Glasgow,
> elsewhere, mesmerism, Mexico, clothes, outsider.

(*C.*) Read:—

[shorthand]

---

* See under ā page 37.

[shorthand text — 7 lines of linear shorthand notation]

(*D.*) Write in Linear Shorthand :—

THE ARMADA.—(i.) A GAME OF BOWLS.

On the 19th of July, 1588, a group of British sailors was gathered on the bowling green at Plymouth Hoe whose peers have never before or since been brought together, even at that favourite meeting place of the heroes of the British Navy. There was Sir Francis Drake, the first English circum-navigator of the globe, the terror of every Spanish coast in either hemisphere ; there was Sir John Hawkins, the rough veteran of many a daring voyage and desperate battle in the African and American seas ; there was Sir Martin Frobisher, one of the first to penetrate the Polar Regions in search of the north-west passage : there was Lord Howard of Effingham, beneath whose leadership the whole fleet had been placed ; and many other fearless mariners whose names have not come down to us. A match at bowls was being played, in which Drake and other high officers of the fleet were taking part, when a vessel was seen running in towards the harbour with all her sails set. Her master, whose name was Fleming, came ashore. He hurried to where the English officers were playing, and said that he had that morning espied the Spanish Fleet off the Cornish Coast. At this exciting news all began to hurry down to the water, and there was shouting for the ships' boats, but Drake coolly insisted that the match should be played out, for " they had time enough to win both the game and the battle." They needed no urging to resume the match, the best and bravest that was ever scored. The winning throw was made, and then they rowed out to their ships and prepared for the great fight, with hearts as light, and nerves as firm, as they had been upon the Hoe Bowling-green.

[*The above passage contains 301 words ; and should be written from dictation in ten minutes.*]

LESSON XII.

COALESCENT CONSONANTS—Continued. (T. D. N. KON OR SHON. W. Y.)

Continuing our examination of the second class of consonant combinations, we come next to the coalescent forms for t and n, which are represented by flat horizontal curves.

[II.] COALESCENT T.

(i.) Coalescent t is represented by a small flat curve having the convex side upwards, thus:— ⌢ . The corresponding long curve ⌢ represents d, thus:— pt, bd, ft, vd, mt, lt, ld, nt, kt, gd, &c., as in:—

reptile. believed.
empty. felt.
child. venting.
checked. aged.

(ii.) These signs are only used at the lower end of consonants, *i.e.*, where t or d is second in a pair. The combinations where t or d is first in a pair are excessively rare, and fall properly under the head of "accidental" combinations. Such words, therefore, as *Whitby, Bidkar, Cadman*, will be treated of under that section; they must *not* be written with coalescent t and d.

(iii.) After , at the end of words, the t hook is turned backwards, thus:— pt, mt, nt, plt, as in—

slipped. tempt.
yclept. recount.

After , at the end of words, the d hook is turned backwards thus:— md, wd, nd, ngd, as in—

reclaimed. seemed.
dawned. harangued.

(iv.) The symbols for s or z, t and d, combine readily with each other, thus:—

axe. act. acts.
thefts. adapts.
tent. tents.
defend. defends.
lapse. lapsed.
against. mainstay.

(v.) Where coalescent t follows coalescent s, combined with a character ending with a hook, the t hook is turned backwards at the end of words (compare iii. above), thus :—

taxed.　　　　fixed.

waltzed.　　　giv'st.

(vi.) Consonant groups like **trst. wd. tt. ss.** &c., which cannot be sounded according to strict rule, are used for syllables by understanding a neutral vowel between the component parts. This enables us to write many common short words in a very abbreviated form, thus :—

trust.　　　　best.

was.　　　　would.

did.　　　　must.

but.　　　　trade.

date.　　　　says.

fences.　　　gravitate.

pressed.　　　first.

could.　　　made.

Exercise XVIII.—(*A.*) Write in Longhand.

(*B.*) Write in Linear Shorthand :—

recollect, flinty, afternoon, halter, script,
adepts, clefts, elapsed, perplexed, adapted,
perched, brandy, tumult, empty, climb,
Windsor, exalts, Ribston, vexed, *but* dexter,
altitude, Cobden, writhed, Ireland, flagrant,
attempts, shields, deemster, reputed, largest,
fault, vaunting, aloft, flaunt, fragrant,
withstand, transact, cam'st, fill'st, *but* Falstaff,
wilderness, identify, averaged, aimed, presentiment,
students, mixed, Anstey, repressed, boasted.

[III.] COALESCENT N.

(i.) Coalescent **n** is represented by a small flat curve having the convex side downwards, thus :—　⌣　The corresponding long curve　⌣
represents the prefix **kon** at the commencement, and the suffix **shon** at the

end of words; medially it is used for either syllable. Thus:— nv, konv, nm, nr, nt, kont, nth, nk, konk, ntr, kontr, nfr, nfl, nkr, koncr, &c., &c., as in—

envy.	convict.
enmity.	Henry.
entire.	contact.
labyrinth.	encaustic.
conquer	introduce.
contrivance.	Lanfranc.
inflate.	Tancred.

It will be seen that at the commencement of a word, ĕ is read before coalescent **n**.

(ii.) These signs are generally required as shown above, only at the *upper* end of consonants, i.e., where the **n** or **kon** is the first in a combination of consonants; occasionally, however, the ‿ , and frequently the ‿ are required at the *lower* end of consonants, usually at the end of words. In this position, when the consonant to which they are attached ends without a hook, the ‿ and ‿ are joined by means of a very short upstroke, which may be traced back up the stem of the character, thus :—

fn, ln, thn, rn, lshon, kshon, pn, tn, mn, ngn, nshon, pshon, as in—

Solney	cognate.
Stepney.	amnesty.
convulsion.	contraction.
option.	invention.

(iii.) At the lower end of consonants, when followed by a curved vowel, ‿ is enlarged in the same way as the terminal hooks of consonant characters, thus :—

Vishnu.	Abner.
Agnostic.	recognise.

It is evident therefore that ‿ cannot be followed by a vowel when used at the lower end of consonants. In such cases (which are most rare) the vowel which follows the **shon** should be omitted, and the words written like -

optional.	functional.

otherwise the alphabetic character for **shon** must be used.

(iv.) Before **p** and **b**, coalescent **n** and **kon** are generally read **m** and kom ; before **f** they are often read **m** and **kom** ; before **th**. **n** is sometimes read **ng**, thus :—

_empowered.
_lamprey.
_compliments.
_Humphrey.
_strength.

_combat.
_timbrel.
_rambling.
_emphasis.
_length.

(v.) The symbols for **n** and **kon** combine readily with those for **s** and **z**, thus :—

_henceforth.
_instruct.
_discontent.

_instance.
_consternation.
_fractions.

Exercise XIX.—(*A.*) Write in Longhand :—

(*B.*) Write in Linear Shorthand :—

Hanley, unfounded, lymphatic, Inman, interested, temperate, strength, evening, Vishnu, optional, emotional, commissioned, constructed, Henry, labyrinth, anchor, antipathy, inconsiderate, recognise, Abner, conversion, tradition, consternation, imitate, rancour, panther, imply, amply, improvident, lamprey, triumph, agnate, ethnography, prescription, vanguard, conversational, instrumental, centralize, intrigue, infringe, ambidextrous, lengthy, hymnal, concord, Walter, compared, attraction, revulsion, revolution, revelation.

(*C.*) Read aloud.

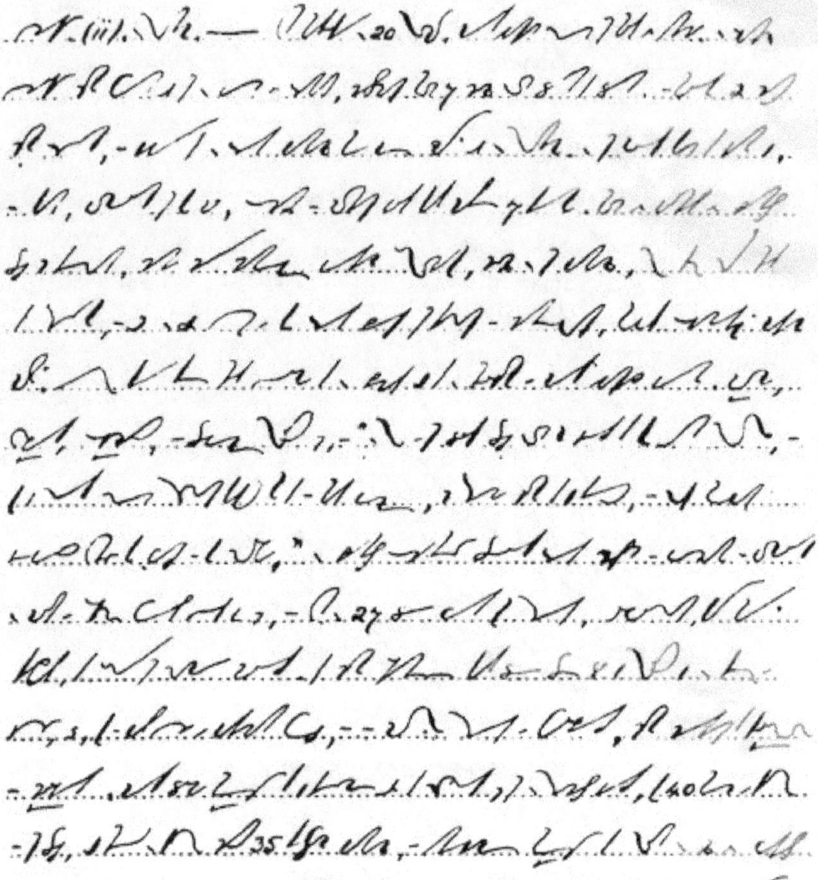

[shorthand text spanning approximately 20 lines]

(*D.*) Write in Linear Shorthand :—

THE ARMADA.—(ii.) THE CHANNEL.

On Saturday, the twentieth of July, Lord Effingham came in sight of the enemy. The Invincible Armada was drawn up in the form of a crescent, measuring some seven miles from horn to horn. A south west wind was blowing, and before it the great vessels sailed slowly up the Channel. The English allowed them to pass by, and then, following in their rear, commenced a running fight that lasted seven days. Some of the finest of the Spanish ships were taken, many more received fatal injuries, whilst the English vessels, which took care not to close, and made the best use of their great speed in tacking and manœuvring, suffered comparatively little loss. Each day added not only to the spirit but to the numbers of Lord Effingham's force. Raleigh. Oxford, Cumberland, and Sheffield joined him, and " the gentlemen of England hired ships from all parts at their own charges, and with one accord came flocking thither as to a set field, where glory was to be attained, and good service performed unto their prince and their country." The Spanish commander showed great wisdom and firmness in following the line of conduct traced out for him, and on the twenty-seventh he brought his fleet unbroken, though sorely distressed, to anchor in Calais roads. It was intended that he should here be joined by the Duke of Parma, who, with a vast army of veteran troops, and a large fleet of transports, was waiting at Dunkerque and Meinport. Lord Henry Seymour had been told off to blockade him in the Flemish ports, with forty sail of Dutch and English ships, but now the Dutch manned thirty-five additional vessels, and enabled Seymour to join the rest of the British fleet, which lay off Calais roads, waiting for an opportunity of attacking their unwieldy foe.

[*The above passage contains* 304 *words, and should be written from dictation in ten minutes.*]

[IV.] COALESCENT W AND Y.

(i.) Coalescent **w** is represented by a small back circle or loop ; it is only required at the lower end of characters, and is thus written :   pw,

✓ fw,   ⌐ mw,   ⌐ lw,   ✓ tw,   ⌐ kw,   ⌐ chw, as in :—

_____ Hepworth.   _____ always.

_____ between.   _____ equal.

_____ question.   _____ anguish.

_____ Edgware.   _____ wigwam.

(ii.) Coalescent **y** is represented by a small downward loop, below and to the right of the character to which it is attached. It is only used at the foot of characters, and is thus written :— ⌐ by, ⌐ my, ⌐ ly, ⌐ ty, ⌐ dy, ⌐ ny, ⌐ ky.

Its most common employment is in the representation with **ĕ** or **ŭ** of the sound **yoo**, which is generally slurred in pronunciation, thus :—

_____ regular.   _____ angularity.

_____ nature.   _____ features.

_____ augury.   _____ salutary.

It is also used as a substitute for **ĭ** and **yoo** followed by another vowel, as in :—

_____ usual.   _____ Miriam.

_____ atheist.   _____ virtuous.

_____ vitiate.   _____ anæmia.

Finally it is used, where necessary, to distinguish between **ōō** and **yōō**, as in :—

_____ ado.   _____ adieu

_____ poor.   _____ pure.

_____ fool.   _____ fuel (*lit.* : fyool).

[*For exercise on Coalescent* **w** *and* **y** *see end of next section.*]

[V.] GROUPS OF COALESCENTS.

When a consonant group is composed entirely of consonants which possess coalescent forms, it is obvious that we require rules to decide which of them is to be represented by its alphabetic character These rules are four in number.

In combinations of consonants all of which have a coalescent form :—

(i.) **kon** and **shon** are always represented by the coalescent form.

(ii.) Except when it would interfere with the above rule, **s** or **z** is always represented by its coalescent form.

(iii.) Except when it would interfere with the above rules, **an**, **en**, and **in** at the commencement of words are represented by the coalescent form.

(iv.) In all other cases the first consonant of such a group is represented by its alphabetic character.

These rules have been most carefully worked out, and must be strictly observed. They cover all possible cases and are sufficient in themselves to remove all doubt, but it may be of some service to the learner if we give their results in a more concrete form. It must be remembered that they apply only to groups of consonants which all possess coalescent forms.

(i.) The general rule is that the first consonant of such a group is represented by the alphabetic character, thus :—

_ _ _ _	rent,	*not*	
	fancy,	*not*	
	Linwood, *not*		

This rule is subject to the following exceptions :—

(ii.) For **an**, **en**, **in**, followed by a consonant at the commencement of words, use the coalescent form of **n**, except in the syllables **ans**, **ens**, **ins**, followed by a vowel ; when **n** is represented by the alphabetic form, thus :—

antiquity,	*not*	
ends,	*not*	
ensnare,	*not*	

But :—

answer,	*not*	
ensue,	*not*	
[Anstey,	*not*	].

(iii.) For **s** use the coalescent form, except in the syllable **kons** followed by a vowel, thus :—

grist,	*not*	
incisive,	*not*	
consternation,	*not*	
sweepstakes,	*not*	
onset,	*not*	

But :—

console,	*not*	
conciliate,	*not*	
reconsider,	*not*	

The two last lessons, on the coalescent consonants, have been long and somewhat difficult perhaps to the learner. Special pains should therefore be taken that they are thoroughly mastered, as it is of the first importance to write the coalescents correctly.

Exercise XX.—(A.) Write in Longhand :—

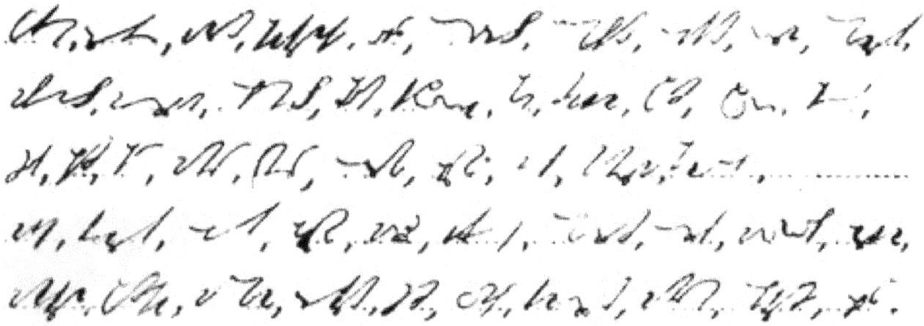

[shorthand/longhand script characters]

(B.) Write in Linear Shorthand :—

traduce, created, valiant, individuality, umpire, conflagration, inconsistent, competent, actual, conciliate, restoration, requiem, consternation, entwine, destructive, anthem, answerable, entrance, instruct, endow, insight, instance, enter, renter, under, congress, impiously, amazed, century, antelope, lapwing, deviate, comfort, amphibious, require, between, consequence, conquest, recognised, enviable, residual, intransitive, reconcile, gradient, incense, promised, delinquent, lantern, constituency, inspire.

LESSON XIII.

ACCIDENTAL CONSONANT COMBINATIONS. (Third Class of Consonant Combinations.)

The third class of consonant combinations, that of accidental combinations, includes all cases which have not been provided for in the three preceding lessons. Such combinations are written with the ordinary alphabetic characters, but, in order to avoid the presence of a liaison, the second is written straight on from the point where the first one terminates, thus :—

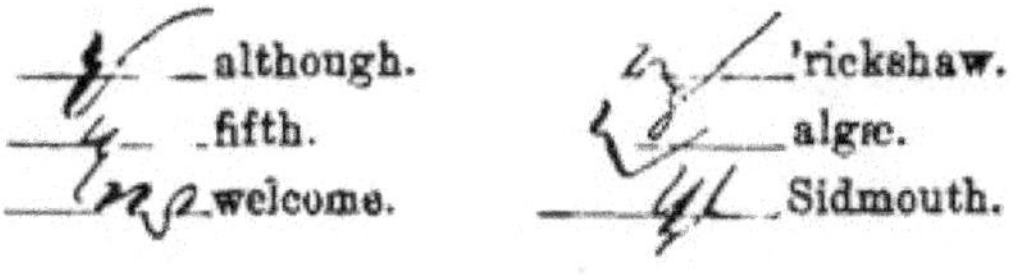

[shorthand characters] —although.
[shorthand characters] .fifth.
[shorthand characters] .welcome.

[shorthand characters] 'rickshaw.
[shorthand characters] .algæ.
[shorthand characters] .Sidmouth.

A very short upstroke (the same which is used to connect ⌣ and ⌣ to characters ending without a hook) is introduced where necessary, *i.e.*, where there is no hook at the junction, thus :—

    —⁄⁄   depth.   *ƒ⁄⁄*   Shorthand.

Exercise XXI.—(*A.*) Write in Longhand :—

[shorthand outlines]

(*B.*) Write in Linear Shorthand :—

     acme, wrathful, owing, Xerxes, Ridpath,
     Shorthand, troubadour, troubadours, dread, **Alcmœna**,
     venture, ringleader, wealthy, depth, bodkin,
     iron, I'm, earn, Tyne, vine,
     Tilbury, almost, self, vulgar, **fifth,**
     see, seer, sir, buoyed, boys,
     Cadmus, backsheesh, width, emotional, doctrine,
     insured, alphabet, wigwam, tipcat, Shakespeare,
     Colepepper, blunder, although, best, promulgation,
     provisioned, **belfry**, instance, subject, remnant.

(*C.*) Read aloud :—

[shorthand outlines]

[shorthand characters]

(*D.*) Write in Linear Shorthand :—

THE ARMADA—(iii.) THE FIRE OF ANTWERP.

The Armada lay off Calais with its largest ships ranged outside "like strong castles fearing no assault, the lesser placed in the middle ward." The English admiral could not attack them there with any hope of victory, so he resorted to a stratagem that had recently been used with terrible success by the Dutch at Antwerp. He prepared eight fireships, small vessels filled with inflammable materials, and sent them at midnight down towards the Spanish lines. The captains in command, whose names were Young and Prouse, effected their hazardous duty and took to their boats. Then suddenly the fireships burst into flame, and bore down upon the Armada with deafening explosions and a glare as of midday. The startled Spaniards were taken completely by surprise, and crying, "The fire, the fire of Antwerp," slipped their cables and fled in all directions. The confusion became terrible. Some dashed for the open sea, some beached their vessels in despair; some of the largest galleons, colliding in the semi-darkness and confusion, drifted helpless and sinking on the rocks; many running foul of the fireships, caught the conflagration and added new danger and horror to the scene. A violent tempest had also now set in, with a furious gale from the south-west. The rain descended in torrents, and the crash of the thunder, and the vivid flashes of the lightning mingled with the glare of blazing galleons and the roar of bursting magazines. So passed that terrible night. A grey and sullen morning broke at last: the flames burned out and the thunder died away. The great Armada, which at sunset had thought itself so invincible, lay scattered all along the coast from Calais to Ostend. Their one thought was of flight, their only prayer for safety; all hope of conquest was now irretrievably destroyed.

[*The above passage contains 302 words, and should be written from dictation in ten minutes.*]

LESSON XIV

NEUTRALIZATION OF VOWELS.

As explained on page 5 of the Introduction, it is unnecessary and inadvisable to express all vowels exactly, provided that their places are shown. When the means of indicating consonant grouping, which have been given in the preceding chapters, are made use of, most medial vowels may be represented by the vowel ŏ, which has been used throughout the system for the neutral vowel. This may be called the Neutralization of Vowels. It will not be found to cause any ambiguity or even hesitation if the consonant groups are correctly written. It will, however, be well for the learner to neutralize at first only unaccented vowels, extending the practice as his experience increases. It is rarely wise to neutralize initial or final vowels, or accented long vowels.

_____ ópposite.   _____ calámity.

_____ cúltivate.   _____ fállacy.

_____ majórity.   _____ résonant.

Exercise XXII.—(*A*) Write in Longhand :—

(*B.*) Write in Linear Shorthand :—

barbarous, persistent, federal, sublime, altogether,
representative, territory, accompanied, developed, recognized,
antagonistic, morality, attributed, liberty, declaration,
family, colony, aspirations, emigrants, political,
amalgamated, difficult, American, separate, promulgated,
supremacy, adventurous, agitated, successive, fatal,
magistrates, protestant, republic, democratic, patriotic,
majority, avocation, nobody, interrogate, contrast,
dominant, symptom, mountainous, decorous, common,
gathered, jealousy, demonstrate, ignorance, rhetoric.

(C.) Read aloud:—

[shorthand characters]

(*D.*) Write in Linear Shorthand :—

THE ARMADA—(iv.) THE RETURN.

Fierce blows the gale, and fast they sail, borne by the following wind,
Quickly they leave the shores of France and the shallow seas behind ;—
Three days, three nights, till the polar lights sprang up in a flick'ring sky,
Three nights, three days, o'er the ocean ways, to the northward still they fly.
Where billows seethe in yeasty foam round Dunnett's home of snows ;
Where mad Cape Wrath is veiled in cloud when the wet south wind blows ;
Where the swift currents to and fro race through the narrow seas,
Betwixt the coasts of rocky Skye and stormy Hebrides ;
Where rise the stately pillars on Staffa's wondrous shore ;
Where sea-birds flock on reef and rock by lonely Skerrivore ;
Where all day long the thwarted tides around the Blaskett's rave ;
Where Connemara's cliffs defy the wild Atlantic wave ;
On, drifting on, from cape to cape, the sport of sea and sky,
Scattered and helpless one by one the reeling galleons fly.
Anon they hear on rocks unseen the breakers roar aloud,
Or catch one glimpse of ragged cliffs through wracks of driving cloud ;
Anon the ruddy beacon light gleams fitful through the spray,
Where the fierce wreckers lay their toils for the heaven-devoted prey.
"Oh, Lady Mother, hear us, and bring us safely home !
Full many a wide estate we give to the Holy Church of Rome !
Full many a noble gift we vow to the Blessed Saints to pay : "—
But the icy blast in fury passed, and swept the prayer away.
The raging heavens answered with the sullen thunder's roll,
And winds sung mournful requiem above each sinking soul ;
Upon the shore, upon the sea, their scattered corpses lay,
And stranded wrecks and tossing spars long marked destruction's way.
Sadly the remnants of the fleet that left so haughtily,
Bare home the tale of the pitiless gale and England's victory ;
And grief and utter horror descend on sunny Spain,
And night and day the people pray for the spirits of the slain.

A. J. C.

[*The above passage contains 334 words, and should be written from* **dictation** *in eleven minutes.*]

CHAPTER V.

The subject of abbreviation is fully dealt with in the second part of the Manual, where it is divided into two parts, Contraction and Elision. Contraction is an exact method, and provides definite contracted forms for common words and syllables. Elision is ambiguous, and indicates what sounds and syllables may be safely omitted. Contraction is the only kind of abbreviation used in the corresponding style, and that only to a limited extent. It may be divided into three classes, contraction of word signs, contraction of prefixes, and contraction of suffixes, and each of these may be again divided into two sub-classes, the first of which is fully, the second only partly, dealt with in the present part. In addition to these forms of contraction, this chapter will introduce the learner to the use of "Phrasing," or the joining together of words and phrases, by which he may save the time otherwise wasted in lifting the pen.

LESSON XV.

WORD SIGNS.

The word signs used in Linear Shorthand may be divided into two classes; first, alphabetic word signs, representing chiefly the common short words—prepositions, articles, adverbs, and auxiliaries—which occur in considerable numbers in every sentence; and secondly, compound word signs, representing technical words and phrases of constant occurrence in the subjects with which shorthand writers chiefly deal. Of these two classes, only the first will be fully dealt with in the present chapter, the use of compound word signs being further developed in the second part of the manual.

(A.) ALPHABETIC WORD SIGNS. (Second List.)

In Lesson V. a list of alphabetic word signs was given, to be committed to memory. The following is a complete list for the corresponding style. Some persons will be unable to learn all of these 165 words at one time, and will do better to introduce a few of them only at first into their writing, and gradually increase the number; but it must be pointed out that the characters are not made to stand arbitrarily for words which they do not suggest, as is the case in many systems. In many instances the sign exactly represents the sound of the word; in all it immediately suggests it. These word signs are formed generally in accordance with the two rules given in Lesson V., to which it will be well to refer again.

ALPHABETIC WORD-SIGNS.

one, special, especial, perhaps, part, or party, past.

be, or been, or by, has been, absent (ce), but, about, best.

for, if, after, first.

very, every, have.

me, or my, or may, am, him, Mr. small, might, am not, or amount, must, most, almost, it must, man, or men, made.

where, or were, or we, was, it was, it was not, always, what, would, it would.

right, rest.

let, all, will, or ill, less. only, last, or lest, well, lord.

to, at, it, still, it is, or its, it is not, time, instance, interest, toward.

do, had, does, it does, did, it did, doubt, does not,

think, or thing, hath, things, thought.

they, or there, or their, although, with, without, within, that, this, there is, this is, then, or than, they'd.

no, or know, when, necessary, since, England.

∠ so, *∪* also, *∠* says, *∠* said.

∠ has, *or* as, *∠* his, *or* is, *∠* has not, *∠* is not.

∪ can, *∪* cannot, *∪* circumstance, *∪* could, *⌐* according, *or* according to, *∪* country, *⌐* contrary, *∠* question.

∪ go, *∪* God, *or* good, *∪* against, *∪* again.

∪ which, *∪* such.

∪ gentlemen, *∪* just.

∫ who, *∫* whose.

∫ your, *or* year, *∫* yours, *or* years, *or* yes, *∫* yet.

∫ shall, she. *∫* wish.

C present, *C* strong, *C* trade.

∪ from, *∪* friend, *∪* through, *∪* throughout.

∖ place, *∖* places, *∖* plaintiff, *∖* able, *∖* ability.

∖ Christ, *∖* great, *∖* client.

• self, *∪* in, *⌐* on, *∩* not, *..* into, *∩* under.

– a, *or* an, *or* and, *–* of, *ˏ* the.

Exercise XXIII.—Make an alphabetical list of the words given in the above table, with their shorthand equivalents. This should be kept for reference when required.

(B.) COMPOUND WORD SIGNS.

There are many common words and phrases which occur so frequently in certain work, that they are commonly represented in longhand by their initials. Such words and phrases are P.S.; B/L.; c/o.; M.S.; &c. These, and other analogous combinations, are represented in Linear Shorthand by compound word signs, formed by joining the initial or principal letters together without intervening vowels. This method is, in fact, a simple development of the alphabetic word signs formed with the help of coalescent consonants. The following list will indicate the manner in which these word signs are formed; it may be added to by the learner if he finds it desirable.

COMPOUND WORD SIGNS.

Account current.	Free on Board.
Account sales.	General Manager.
Balance Sheet.	Manuscript.
Bankrupt (cy).	Market Price.
Bill of Exchange.	Months after date.
Bill of Lading.	Per annum.
Bill of Sale.	Per cent.
Board of Trade.	Per Proc.
Care of.	Postscript.
Chamber of Commerce.	Post Office Order.
Chancellor of the Exchequer.	Promissory Note.
Defendant.	Pro tem.
Discount.	

Exercise XXIV.—(*A.*) Write in Longhand :—

(*B.*) Write in Linear Shorthand :—

according-to, friend, Chancellor-of-the-Exchequer, party, country, Lord, God, Christ, hath, rest, special, ability, client, plaintiff, defendant,

amount, interest, discount, per cent., per annum,
trade, bill-of-exchange, months-after-date, within, time,
C/o., Mr., well, your, P.S.,
place, whose, per proc., B L., perhaps,
they-would, about, cannot, free-on-board, after,
small, bill-of-sale, yes, let, bankrupt,
always, although, also, almost, against,
shall, only, pro tem, still, under,
promissory-note, thought, instance, it would, when.

(*C*)--Read aloud :—

(*D.*)—Write in Linear Shorthand :—

PARTY POLITICS.

My first duty *this* evening, *gentlemen, is,* I *think, to let* you *know* how *very* deeply I *am* touched *by the most friendly* feelings *towards my self which* you *have* testified to-night, *after my* long *absence from* amongst you ; *and also, although* I *could wish that they might have been less ill* deserved *than they* are, *by the* kind *things that were said just* now *by Lord* —— *and Mr.* —— *of such small* services *as* I *may have been able to do* you *from time to time.* I *am-not so* foolish *as to* suppose *that all this good will on your part is* extended *only to my self as an* individual. I recognise, *on the contrary, that it-is* rather *a* mark *of your special* approval *of the* principles *of the great party which it-was always, and still is, my* chiefest boast *to* represent *in this* constituency.

Well, gentlemen, since the day *when* I *had last the* honour *of* meeting you *in this place, circumstances have greatly* altered. *Then, according-to* our opponents, *we were the* miserable remnants *of a thoroughly* defeated army. *It-is-not necessary at the present time to go again into the* causes *which had made their* victory *so* complete. *No doubt* each *one of* us *has his* own *thoughts about it, and perhaps* he *cannot rightly* complain *if every one* else *does-not* agree *with* him. *For years, it-must be* allowed, *the country* districts, *where in the past* our *party was so strong, have been against* us *almost to a man. But at last there-is, without* any *doubt,* clear indication *throughout all parts of England, that* those *who can* see *what their best interests* really demand, *have made* up *their* minds *to put an* end once *and for all to the* system *of* disaster *and* misrule *under which we have been* groaning *for so* long.

[*The above passage contains* 313 *words, of which* 79 *are fully written, and the remainder, which are printed in italics, are represented by* 224 *word signs and groups. About* 130 *different word signs are made* use of. *The passage should be written in ten minutes.*]

LESSON XVI.

PREFIXES.

There is in English a very large class of derivative words formed from verbs and nouns by the addition of *prefixes,* which modify or amplify their meaning. These prefixes, which are for the most part Latin prepositions, form a most convenient means of abbreviation, since the use of a short form for the prefix abbreviates a large number of words.

The prefixes used in Linear Shorthand may be divided into two classes— vowel prefixes and consonant prefixes. Of these two classes, as has been already indicated, only the first is fully treated in the corresponding style, the second being further developed in the second part of the manual.

(A.) VOWEL PREFIXES.

(i.) It was said in **Lesson VIII.** that the liaisons used to indicate initial vowels commenced on the line. Vowel strokes commencing above or below the line are used to represent prefixes commencing with the vowel indicated by the curvature of the liaison and the position of the following consonant.

It was said in the same Lesson that the stroke for initial ă and I was always omitted. These strokes are also used for Vowel Prefixes.

Exercise XXV.—Copy out the following table till it can be written accurately from memory.

Commencement of Vowel Stroke.	Prefixes implied by the vowel positions.					
	ă	ĕ	I	ŏ	ŭ	ȯȯ
Above the line	ab-			of-		
On the line	ad-		ill-			
Below the line	anti- } ante- }	ex-	inter- } intro- }	ob-	un-	uni- } uni- }

It will be seen that the word *of* is treated as a prefix, an arrangement which will be found extremely convenient in practice.

(ii.) The above table shows that the initial vowel stroke for ă is read *ab*, *ad*, or *ante-* (*anti-*), according as it begins above, on, or below the line, thus :—

_____ avocation.
_____ abdicate.
_____ advocate.
_____ antedate.

The other prefixes are formed on the same principle, thus :—

____ . expense.   ____ ill-paid.
____ intermediate. - ____ introduce.
____ objection.   ____ of-course.
____ uncertain.   ____ universal.
____ antidote ; *so also* ____ antagonist.

(iii.) It will be noticed that six of these prefixes end in a vowel, and that seven of them end in a consonant. In the case of these last, it is impossible to show a vowel between the prefix and the root of the word, and. therefore, in words beginning with *ab-*, *ad-*, *ex-*, *ill-*, *of-*, *ob-*, *un-*, followed by a vowel, either the vowel must be dropped or the prefix written alphabetically. Short vowels are generally dropped; when the vowel is long, the prefix is written alphabetically. Thus :—

_↯_ abolition (*lit.* : ablition).

_↯_ abandon (*lit.* : abndon).

_↯_ adequate.

_↯_ existence.

_↯_ unenvied.

_↯_ of-another.

_↯_ ill-affected.

_↯_ excessive.

_↯_ unavoidable.

_↯_ of-imagination.

But :—

_↯_ abuse.

_↯_ ill-ordered.

_↯_ unequal.

_↯_ of-ordinary.

_↯_ adorn.

_↯_ exuberance.

_↯_ unusual.

_↯_ of-evil.

(iv.) As a general rule, **these** and other prefixes should not be used for syllables which contain the same sounds, but are not really prefixes, thus :—

_↯_ abreast,	*not*	_↯_	
_↯_ adrift,	*not*	_↯_	
_↯_ antiquary,	*not*	_↯_	
_↯_ intrepid,	*not*	_↯_	
_↯_ interminable,	*not*	_↯_	
_↯_ obedience	*not*	_↯_	

This rule need not trouble **the** learner, or cause him to hesitate as to whether any such syllable is a prefix or not. If he writes it in the way that seems most natural it will nearly always be right.

(B.) CONSONANT PREFIXES.

Consonant **prefixes** are, [with the exception of *discon-*,] indicated by their initial consonants written disjoined. As has been said, the full development of consonant prefixes is dealt with in the second part of the manual. There are, however, five consonant prefixes which are so common that they are included in the corresponding style. They must be committed to memory and introduced forthwith into the writing.

_/_ = dis-, *or* des-	*as in*	_↯_ , _↯_
_~_ = discon-	*as in*	_↯_ , _↯_
_z_ = mis-	*as in*	_↯_ , _↯_
_l_ = sub-	*as in*	_↯_ , _↯_
_c_ = self-	*as in*	_↯_ , _↯_

Exercise XXVI.—(*A.*) Write in Longhand :—

_↯_

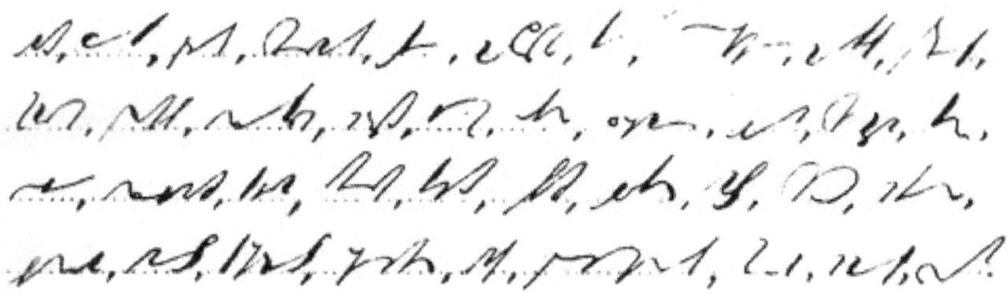

[shorthand text]

(B.) Write in Linear Shorthand :—

obsolete, abject, subsequently, ill-sounding, anticipate,
misaimed, intervene, of-right, unflinching, abstract,
self-evident, ill-starred, explain, disagreeable, of-egress,
abide, adroit, ill-affected, disconsolate, extreme,
advent, abroad, execrate, undulate, unduly,
illustrious, adore, discontinue, ill-ordered, exclude,
subaltern, antagonist, exordium, miscreant, abuse,
of-attack, self-impelled, uniform, unusual, addict,
obey, exuberance, despoil, abandon, adamant,
existence, unavoidable, abolition, unearned, mistake,
introspective, adoration, disintegration, interminable,
admit, excommunicate, suburb, mislead, unicorn.

(C.) Read aloud :—

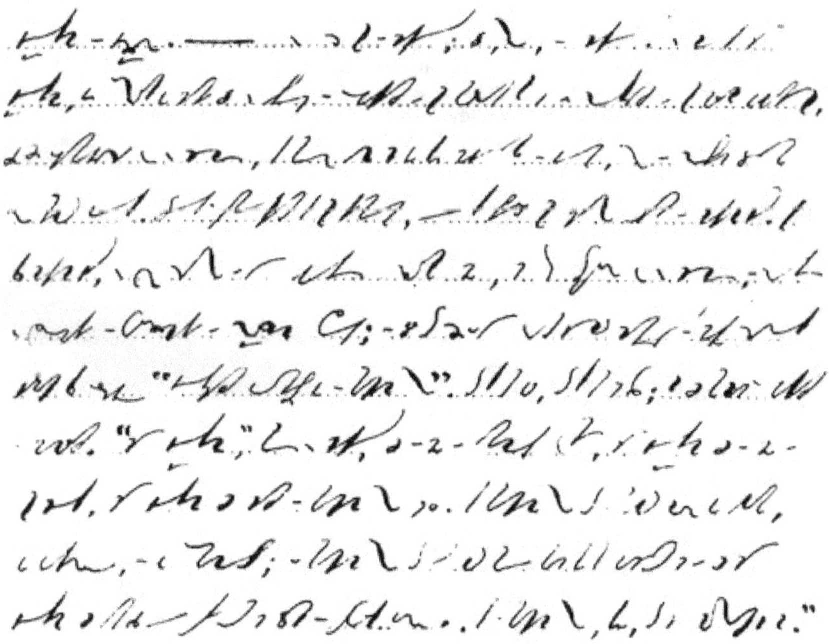

[shorthand text]

[shorthand text]

(*D.*) Write in Linear Shorthand:—

BARDELL *v.* PICKWICK.

The plaintiff is a widow: yes, gentlemen, a widow. The late Mr. Bardell, after enjoying for many years the esteem and confidence of his sovereign as one of the guardians of his royal revenues, passed almost imperceptibly from the world, to seek else where for that repose and peace, which a custom house can never afford. Shortly antecedent to his decease, he had stamped his image upon a little boy. With this little boy, the unique pledge of her ill-fated exciseman, my client shrank from the world, and courted the obscurity and tranquillity of Goswell Street; and here she placed in her front parlour window a written placard bearing this inscription: "Apartments furnished for a single gentleman." She had no fear, she had no mistrust; all was sublime confidence and reliance. "Mr. Bardell," said the widow, "was a man of absolute honour, Mr. Bardell was a man of his word, Mr. Bardell was once a single gentleman himself. To single gentlemen shall I ever look for advice, for protection, and for consolation; in single gentlemen shall I ever see some thing to remind me of what Mr. Bardell was when he first obtained my young and untried affections. To a single gentleman, then, shall my lodgings be let." Animated by this beautiful and touching impulse, among the best impulses of our distorted nature, gentlemen, the lonely and desolate widow dried her tears, furnished her first floor, caught her innocent boy to her maternal bosom, and put the bill up in her front parlour window. Did it remain there long? No, gentlemen; an interval of three days only had elapsed, three days, gentlemen, when a being, erect upon two legs, and bearing all the outward semblance of a man, and not of a monster, knocked at the door of Mrs. Bardell's house. That man was Pickwick—Pickwick, the defendant.

[*The above passage contains* 308 *words and should be written from dictation in ten minutes.*]

LESSON XVII

SUFFIXES.

There are many common terminations used in English, which, like prefixes, form a means of abbreviation which should not be lost sight of in a Shorthand system. In Linear Shorthand they may be divided into two classes, joined and disjoined suffixes, of which the former only are fully used in the corresponding style, the latter being further developed in the second part of the Manual.

(*A.*) JOINED SUFFIXES.

The following consonants and double-consonants, which cannot occur in their usual sense at the end of words, are used in that position to represent certain suffixes nearly resembling them in sound. They are connected according to the rules of vowel indication, and, in the absence of vowels, may be written disjoined when convenient.

⟩ -ward, *or* -wood.

ʃ -head, *or* -hood.

⟩ -ple, *or* -pal.

⟩ -ble, ⟩ -bility, ∨ -bly, ⟩ -bles.

⟩ -fle, *or* -ful, ⟩ -fully, ⟩ -fulness.

⟩ -vle, *or* -vel, *or* -vil, ⟩ -volent.

⟩ -graph, *or* -gram, ∨ -graphy, ⟩ -graphed.

⟩ -cle, *or* -cal, *or* -cule.

As :—

forward.	homeward.
Fleetwood.	hardihood.
childhood.	Godhead.
principle, *or* principal.	
disciple.	ample.
honourable.	noble.
sensibility.	humbly.
enables.	trifle.
hopefully.	thoughtfulness.
revel.	devil.
malevolent.	telegraph.
diagram.	phonography.
heliographed.	icicle.

These suffixes should be committed to memory at once, and introduced forthwith into the writing. They are almost alphabetic, and demand but a slight effort of the memory.

(B.) DISJOINED SUFFIXES.

The following suffixes are indicated by writing their principal consonants disjoined close to the end of the word. Position on the line only is recognised, and vowels cannot be shown at the commencement of the suffix; when present they are written at the end of the main body of the word. The full development of disjoined suffixes is dealt with in the second part of the Manual. Five of them, however, are so common that they are included in the corresponding style. They must be committed to memory, and introduced forthwith into the writing.

___	= -less,	*as in*	
	= -lessness,	*as in*	
	= -ment,	*as in*	
	= -ness,	*as in*	
•	= -self, *or* selves,	*as in*	

Exercise XXVII.—(*A.*) Write in Longhand :—

(*B.*) Write in Linear Shorthand :—
rival, peerless, responsibility, fountain-head, couple,
Norwood, cannibal, heliograph, malevolent, earthward,
edible, abruptly, myself, obscurity, Furnival's,
ill-natured, municipal, liability, of-malice, united,
helpful, thoughtless, bicycle, kilogramme, novel,
unequalled, Minehead, joyfulness, joylessness, tachygraphy,
antipathy, volubility, Roundhead, unstable, wayward,
tackle, heedlessness, motherhood, sinfully, ample,
outward, amiably, business, adduce, autograph,
hitherward, justifiable, tentacle, introspection, themselves,
of-manhood, firmament, stifle, Heywood, dryness,
exhaustive, government, ripple, carnival, childhood.

(*C.*) Read aloud :—

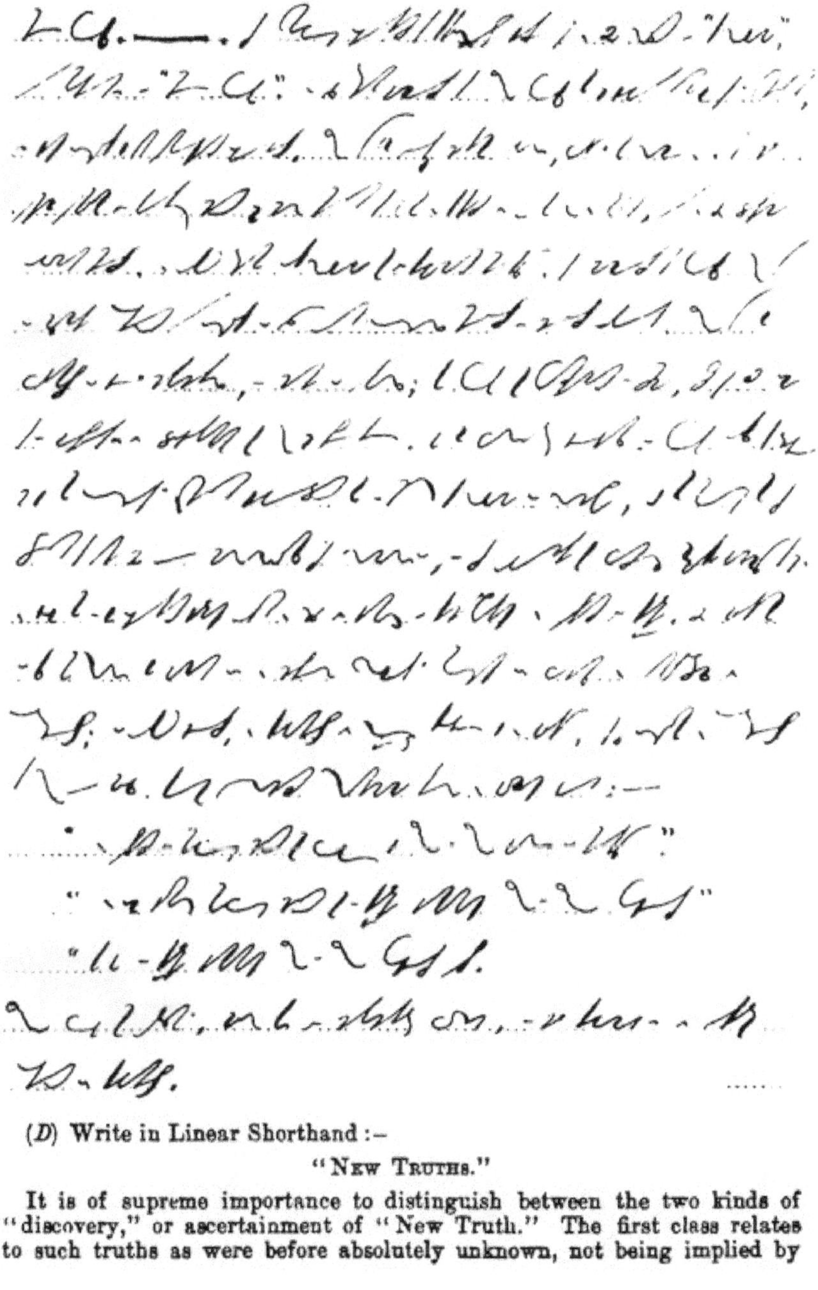

(*D*) Write in Linear Shorthand :—

"New Truths."

It is of supreme importance to distinguish between the two kinds of "discovery," or ascertainment of "New Truth." The first class relates to such truths as were before absolutely unknown, not being implied by any antecedent information. Such are all outward matters of fact, properly so called. No mere internal exercise of a thoughtful mind will make known to us the distance of the earth from the sun, or the past history of foreign nations. The other class of discovery is of a different nature. It relates to truths which are in reality contained or implied in our own complex notions and modes of thought. Such are all propositions in pure mathematics, and many in ethics; their truth is transparent and necessary, since it amounts only to a restatement of the hypothesis with which we started. For all practical purposes a Truth of this description may be as completely unknown beforehand as a novel discovery in chemistry, but as soon as it is shown to any man he recognises its correctness, and its uniformity with principles already familiar to him.

The above has a very important bearing upon the question of the possibility of demonstrating the existence of a Deity. Most writers on this subject have fallen into the mistake of covertly assuming in the premisses the essentials of the conclusion; in other words, the definition of the Godhead adopted by the writer, itself implies the conclusion at which he arrives. These arguments generally take the following form :—

" The existence of a supreme mind is proved by such and such facts in nature.
" The only possible supreme mind is a deity possessing such and such attributes.
" Therefore a deity possessing such and such attributes exists."

Such proof is obviously, like that of a mathematical problem, a mere development of the ideas contained in the definition.

[*The above passage contains* 304 *words, and should be written in ten minutes.*]

LESSON XVIII.

PHRASING.

In rapid writing it is natural to run from word to word without lifting the pen. Each lift represents a small amount of time, and therefore a free use of phrasing is a very real aid to speed. In Longhand it is nearly always possible to connect words without obscuring the sense, but in most Shorthand systems phrasing is so employed that the phrases are almost unintelligible in themselves, and the learner is obliged to commit to memory long lists of those which he may safely use. This is due chiefly to the admission into phrases of vowel-less abbreviations and arbitrary word signs, without regard to the *sound* of the resulting combination. The leading rule for phrasing in Linear Shorthand is simple and comprehensive, and is based upon the fact that, in speaking, the break between two words is no greater than that between two syllables. It is that the *sound* of a phrase must be written as though it were a word, strict attention being paid to the rules of vowel indication and of consonant grouping. The result of this rule is that phrasing may be used as freely as in Longhand without the slightest hesitation either in reading or writing. The word signs have been so selected (with a view to their use in phrases) that the majority of them may be introduced, when desired, without change; the modification which is needed by others is generally only the addition of an upstroke when they occur at the end of a phrase.

The rules, then, which govern the use of phrasing in Linear Shorthand are as follows :—

(i.) Groups of words may be treated as a single word, and written without lifting the pen, provided that the rules of vowel indication and of consonant grouping be observed. Thus :—

as-it-seems.
at-any-rate.
if-I-may-say-so.
you-may-be-told.
in-spite.
of-course-it-is.
you-never-saw.

(ii.) Of two consecutive vowels, the weaker may be omitted, and, generally, the colloquial pronunciation followed. / is used for " the," and the neutral vowel for " a." Thus :—

so-far-as-I-see.
as-he-has-told-us.
I-am-amazed.
it-is-easy-to-say-so.
the-other-day.
it-is-a-fact.

(iii.) Word signs may be admitted into phrases when they represent nearly the full sound of the word, but not otherwise. Thus :—

about-a-year-ago.
it-is-always-a-very-good.
that-there-will-be.
perhaps-it-will-be-as-well.
have-you-been-asked.

But :—

only-a-*small*-amount.
in-my-*absence*.
necessary-evil.

(iv.) When word signs in which other sounds follow the consonant by which they are represented, occur at the end of a phrase, the last vowel of the word sign must be written, if not exactly, at least approximately.

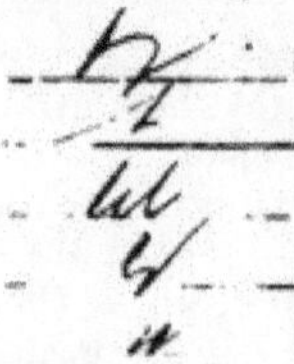

do-you-know.
ought-to.
the-very-thing.
as-it-were.
may-be.

Exercise XXVIII.—(*A.*) Write in Longhand :—

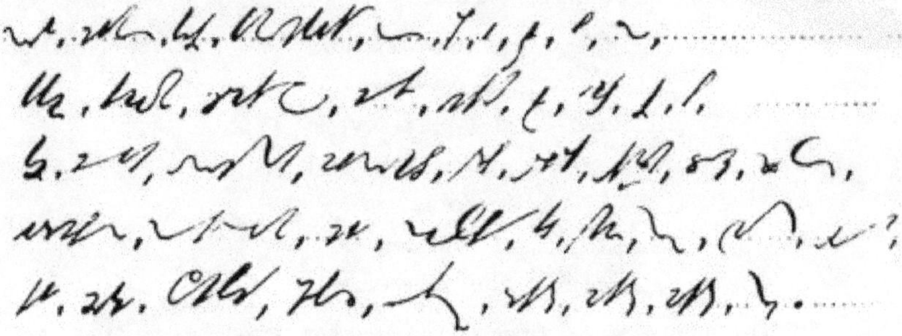

[shorthand content]

(*B.*) Write in Linear Shorthand :—

it-cannot-have, it-may-be-said, that-it-is-not, the-sooner-the-better,
friend, instance, one, balance-sheet, has-been, circumstance,
that-there-will-not, deliver-us, it-was-only-too-true, we-ought-to,
intermittent, Board-of-Trade, abolition, it-does-not, still,
as-it-was-not, more-often, I-cannot-imagine, respect-his-wishes,
antichrist, uninspired, Unitarian, homeward, quadruple,
a-very-great-mistake, free-to-confess, it-may-be, give-us-this-day,
adept, exonerate, Chamber-of-Commerce, unravel, uniform,
it-has-been, it-must-have-been, strong-as-they-were, in-another-place,
artfulness, critical, reticule, ridicule, general-manager.

(*C.*) Read aloud :—

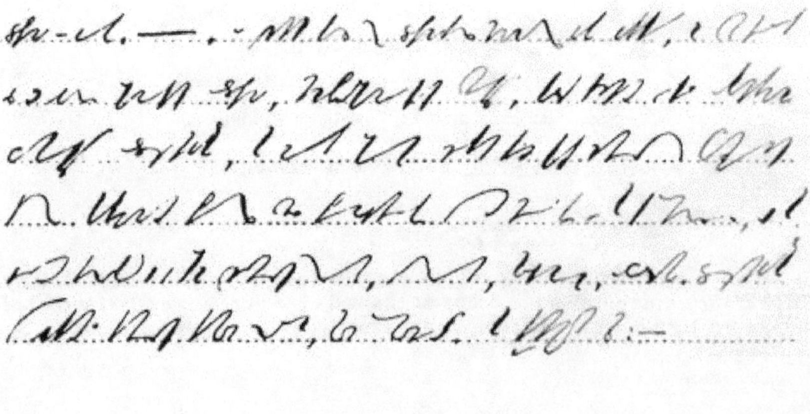

[shorthand content]

(D.) Write in Linear Shorthand : —

HISTORY AND FAITH.

In examining the-place which history-takes in-a-religious faith for-to-day, let us note first what effect in-the-reading of-history, no-less-than in-the-reading of-nature, the-great dominant idea of-the-natural procedure of-human-affairs, has wrought in-these modern times with-its wonderful transforming touch. The-different stages of-man's story-upon-earth are-not now thought of as disconnected, but as bound together by one endless unwinding chain, a-chain, as-I-believe, of-progress. Human-affairs are evidently tending to-some goal, some consummation. As Tennyson says : —

> " Yet-I doubt not thro' the-ages, one increasing purpose runs,
> And-the thoughts of-men-are widened with-the process of-the-suns."

Now for-the-purposes of-a religious faith this conception is of-paramount importance ; because it-is self-evident that-if-it means-anything it means that God is-in-history. And it means also that God never contradicts Himself, that-He-is-present in-what, for-the-purposes of-historical convenience, we-call one age, as-He-is-present in-another. Never and no-where has-He-left-Himself without witness, and to-the-religious mind this great-fact will rightly become of-truly Divine significance. In-the-same-way, such-a-mind will-also dwell with gratitude upon the-different stages of-the-story chronicled by history. It-will-see that-every-nation which-has-ever-been or-now is, has-had-its part to-play in-the universal progress towards that consummation to-which humanity as-a-whole is tending. It-will-also-see in-the-ceaselessly ascending types of-character admired by-different ages and nations, a-happy-progress towards the-divine-ideal of-man, in-whose ultimate realization all lower ideals have played their parts or are playing them now.

[*The above passage contains* 307 *words, and should be written in ten minutes.*]

We have now finished the exposition of the corresponding style of Linear Shorthand, and the learner should be able to write any passage correctly at a speed of about 30 words a minute, and read it without hesitation. It is easy to increase this speed by dictation-practice to one of 80 or 100 words a minute, without using any other abbreviations than those already employed. For the purpose of practice the learner should repeat the exercises in the last two chapters, and especially the long connected passages, and the examples given below as Chapter VI., until he can write them correctly and legibly at 80 words a minute. He should also take in the weekly numbers of the "Pickwick Papers," which are being issued in Linear Shorthand as explained in the Preface. If he requires a greater speed than 100 words a minute (which is, however, sufficient for most office work), he can with the help of the Second Part of the Manual, which deals with Reporting, attain to any speed he desires. The acquirement of speed is merely a question of practice from dictation, and a certain time should be set apart for this purpose every day, and be strictly adhered to. In conclusion the author would add that he is always more than willing to answer questions about the system, and to give the learner any assistance of which he may stand in need.

CHAPTER VI.

EXAMPLES.

I.—COMMERCIAL CORRESPONDENCE.

[The letters given below are borrowed, for the most part, from " Commercial Correspondence," by P. L. Simmonds, by permission of the Publishers, Messrs. Routledge & Sons.]

The principal duty of a shorthand clerk is to write from dictation business letters, reports, instructions to counsel, etc., and to transcribe them carefully into longhand. In most cases a speed of 80 or 100 words a minute is amply sufficient for this work; but any nervousness or anxiety on the part of the writer will throw him hopelessly behind. It is, therefore, very necessary that he should take special practice in the particular class of work in which he will be engaged. If he is already engaged in office work, by far the best plan is for him to copy, on every available opportunity, the correspondence in the Letter Books of the office. If, however, the learner is not yet so employed, he should make a practice of copying the letters contained in such books as " Simmond's Commercial Correspondence," referred to above, or the articles in specialized journals like the *Shipping Gazette*, the *Economist*, the *Law Journal*, etc.

The eight examples which follow are supposed to be dictated one morning to a shorthand clerk for transcription. The right-hand side shows his notes, the left-hand side the letters as they should be transcribed :—

LONDON, 5TH MAY, 1896.

MESSRS. WALEY AND WRIGHT,
    ADELAIDE.

GENTLEMEN,

    Mr. Rawlings, of this Company, is about to visit the principal cities of Australia for the purpose of extending our business in your country. He sails either by the present or next mail steamer from Southampton. Any information you can afford him, or introduction to houses in our line of business which you can give him, we shall duly esteem.

    Although well supplied with funds, should he stand in need at any time of money, we will thank you to accommodate him on our account to the extent of £700 or £800, drawing upon us at a short date for your advances. Mr. Rawlings bears a letter of introduction from our house, and we append his signature at foot for your information.

        Yours faithfully,

            ROGERS & WAINWRIGHT, LTD.,

(Mr. Rawling's signature.)   p.p.

                        Manager.

LONDON, 5TH MAY, 1896.

MY DEAR RAWLINGS,

    I enclose the letter of Introduction and Credit, addressed to Messrs. Waley & Wright, as you suggested. Kindly add your signature at foot and return it to me.

    I have consulted Mr. Wainwright with regard to the suggestion which you made on Thursday for the establishment of a branch agency in Adelaide. He will be glad to hear your views when you have considered the matter on the spot ; but on the whole he is not at present favourable to the scheme. I should suggest that you leave this matter for personal discussion on your return to England. You know already what are my own views on the subject.

    Wishing you the best of voyages, and a successful journey,

                        I remain,

Encl. : Letter.   Yours very truly,

    H. J. Rawlings, Esq.,
        c/o Rogers & Wainwright, Ltd.,
           Southampton.

LONDON, 5TH MAY, 1896.

W. H. WAINWRIGHT, ESQ.,
    COLMORE PARK,
        FORDHAM,
           SURREY.

DEAR SIR,

    I have written to Mr. Rawlings in accordance with your instructions, asking him to look thoroughly into the question of establishing a branch in Adelaide, when he is in that city, and suggesting at the same time that he should leave the matter for personal discussion on his return.

Waley & Wright.

Enclosed I beg to hand you abstract of the April sales, which are, I am pleased to say, as satisfactory as usual. You will observe that the percentage of increase for the last month is not quite equal to that for the half-year. I do not, however, take that as an indication that we are reaching the limit of possible increases; I consider that it is amply accounted for by the holiday season falling a week earlier than last year; and I expect that we shall feel the benefit of this to an equal extent during the current week.

I remain,

Encl : Abstract Sales.   Yours obediently,

———————

LONDON, 5TH MAY, 1896.

MESSRS. ROGERS & WAINWRIGHT, LTD.,
        MANCHESTER.

DEAR SIRS,

I enclose for your perusal a letter received by us this morning from Messrs. Simpson & Jones of your city, with copy of my reply. Please let Mr. Charlesworth take an early opportunity of calling upon them, and if he thinks well, let him bring Mr. Jones up to town. I could see him any morning next week except Wednesday.

Encl : Letter.   Yours faithfully,
    Press copy.

———————

LONDON, 5TH MAY, 1896.

MESSRS. SIMPSON & JONES,
        MANCHESTER.

DEAR SIRS,

We are in receipt of your favour of yesterday's date, and beg to thank you for the proposal contained therein. We have instructed our Manchester agent to call upon you at an early opportunity and discuss the matter. If after seeing him you think it desirable that you should meet us in London, we shall be happy to fix an hour next week to suit your convenience.

Yours faithfully,
        ROGERS & WAINWRIGHT, LTD.,
    p.p.
                                        Manager.

———————

LONDON, 5TH MAY, 1896.

MY DEAR GREAVES,

Very many thanks for the Commercial Travellers' School tickets. If I can at any time assist you in a similar manner, please count upon me.

Yours very truly,

Robert F. Greaves, Esq.,
    147, Queen Victoria Street, E.C.

Simpson & Jones.

R. F. Greaves.

LONDON, 5TH MAY, 1896.

MESSRS. WILKINS, COLLINS & Co.,
  BOMBAY.

DEAR SIRS,

   We are in receipt of yours of the 3rd ult. covering first of exchange for £34 11s. 6d., for which we thank you. We have not since had the pleasure of hearing from you.

   We regret to hear that your market continues depressed; but we trust that the recent improvement in the Exchange will have a reassuring effect.

   Annexed please find duplicate invoices of the goods indented for, which we hope will be duly taken up on arrival. The B/L, per "Sea Star," was sent you last mail.

Encl: Invoices.

                           Yours faithfully,
                              ROGERS & WAINWRIGHT, LTD.,
                                 p.p.
                                       Manager.

LONDON, 5TH MAY, 1896.

MESSRS. JOHNSTONE & EVERITT,
  BUENOS AYRES.

DEAR SIRS,

   We beg to acknowledge receipt of your favour, last to hand, dated the 24th March, and note contents, which are satisfactory.

   We shall be obliged by your closing as speedily as possible the balance of old shipments such as the following :—

   No. 4 per "Parilla"; No. 9 per "Maria"; and No. 15 per "Ignis Fatuus."

   We should also esteem it a favour if you would send us per return mail an account current to date, to see that our books agree. We shall be glad to have the list furnished by your inland customers as a guide to our shipments to Paraguay.

   Annexed we beg to hand you B/L and Invoice of shipment going forward per 'Witch of the Seas,' which we trust will reach you promptly, and meet with a ready sale.

Encl: B/L.
     Invoice.

                           We remain,
                              Yours faithfully,

                              ROGERS & WAINWRIGHT, LTD.,
                                 p.p.
                                       Manager.

II.—LECTURE NOTES.

   In taking notes of Lectures it is not at all necessary, and indeed it is very inadvisable, to record verbatim the utterances of the speaker. The object of taking such notes is twofold: first, to impress the subject upon the mind during the delivery of the lecture; and secondly, to provide means for reviving the impressions as rapidly as possible when required.

Wilkins Collins :‑

Johnstone & Everitt.

II.

With this view lecture notes should take the shape of an analysis, giving
in a form which may readily catch the eye, the principal heads of the
lecture, and the main features of the illustrations and explanations by
which they are developed. Of course a good deal of practice is necessary
before this can be properly done, and it is not possible to teach the art in a
book devoted to shorthand. It may, however, be of considerable service to
many learners of Linear Shorthand if we give as an instance a portion of
a lecture, first in the form in which it is delivered, and then in the form
in which it should be recorded in the notes of the student. The principle
here illustrated is applicable also to the recording of sermons and political
speeches, when, as is usually the case, a verbatim note is not required.
Notes arranged in this manner can easily be converted into a very full
report, and do not generally demand a speed of more than 50 or 60 words
a minute.

The matter given below is, for the most part, selected and adapted from
Mr. Huxley's well-known "Lectures to Working Men," on the "Causes
of the Phenomena of Organic Nature." It is here presented in a much
altered and abbreviated form, and readers who take an interest in the
subject should refer to the original, which may be found in a volume of
Mr. Huxley's Essays, entitled "Darwiniana," and published by Messrs.
Macmillan & Co., by whose permission the following passages appear.

THE DARWINIAN THEORY.

. . . . What, then, is Mr. Darwin's hypothesis? As I
apprehend it—for I have put it into a shape more convenient for common
purposes than I could find it in his book—as I apprehend it, I say, it is,
that all the phenomena of organic nature, past and present, result from,
or are caused by, the inter-action of certain properties of organic matter
which are called "Atavism" and "Variability," with the Conditions of
Existence. In other words, given the existence of organic—that is living
—matter, its tendency to Atavism and Variability; and lastly, given the
conditions of existence by which organic matter is surrounded, that these
put together are the causes of the Present and of the Past conditions of
Organic Nature. It will be necessary to examine in some detail these
three Causes, and I will write their names on the black-board to impress
them upon your memories—Atavism, Variability, and the Conditions
of Existence. Now, first, let us look at that principle or property
of Organic matter which I have called Atavism. This word,
which means much the same thing as Heredity, comes from the Latin
word *atavus*, an ancestor, and it means the tendency to revert to the
ancestral type. It is a matter of perfectly common experience that the
tendency on the part of the offspring always is, speaking broadly, to
reproduce the form of the parents. The proverb has it that the thistle
does not bring forth grapes; so among ourselves, there is always a likeness,
more or less marked and distinct, between children and their parents,
between cousins, between members of the same family, and even, to a
certain extent, between members of the same country. That is a matter
of familiar and ordinary observation. We notice the same thing occurring
in the cases of the domestic animals—dogs, for instance, and their offspring, which often exhibit curious markings or traits of character which

I. (salt).

were possessed by the parents. In all cases of propagation and perpetuation, whether sexual or non-sexual, there seems to be this tendency in the offspring to take the characters of the parental organisms.

This atavism, then, is one of the most marked and striking tendencies of organic beings; but, side by side with this hereditary tendency, there is an equally distinct and remarkable tendency to variation. The tendency to reproduce the original stock has, as it were, its limits, and side by side with it there is a tendency for the individual to vary in one direction or another; as if there were opposing powers working upon the organism, one tending to take it in a straight line, and the other tending to make it diverge, first to one side and then to the other. The reasons for this tendency to vary, which seems to run across and across the tendency to repeat the parental type, are very obscure, and in the present state of our knowledge I am not sure whether it is worth while to devote much time to them. It is sufficient to know that this tendency is present in all kinds of reproduction. It may, however, be as well to point out that external conditions have something to do with it. For instance, if you take two identical plants and transplant them, one to a richer and the other to a poorer soil, you will probably find that the offspring of the first will vary from the parent in the direction of greater luxuriance, and that the offspring of the other will vary in the opposite direction. By far the greater number, however, of variations, are what we are compelled to call spontaneous variations, by which we mean that we do not know by what they are caused.

Now let us go back to Atavism, to the hereditary tendency I spoke of. What will become of a variation if you breed from it? Well, one very remarkable case has been recorded by Réaumier. It is that of a Maltese—Gratio Kelleia was his name—who was born with six fingers on each hand and six toes on each foot. That is a very extraordinary variation, and we do not know what caused it, so we call it spontaneous. But Gratio Kelleia married; he married a perfectly normal woman, with the fashionable number of fingers and toes, and what was the result? His eldest son had twelve fingers and twelve toes like his father, and three of this man's children were similarly deformed. The next son and the daughter had malformed, though not double, thumbs, and two of their children had twelve fingers and toes, two were partly deformed and four were normal. The remaining son was normal, as also were all his children. Now in this case, in the two generations, out of about twenty persons, six were deformed just like their father or grandfather, four were partially deformed in the same direction, and ten were normal. Now suppose there had been someone who could have played the part of a "fancier" in this case, and "selected" the deformed cousins and mated them, as the pigeon-fancier does in breeding to a certain point, you cannot, I think, have any doubt that he could have produced from this single variation an entire breed of men who should all possess six fingers on each hand and six toes on each foot. To show you that this supposition is not an unreasonable one, I will give you one other instance. In the year 1791, there was born in Massachusetts a very singularly formed ram. It had a very long body, and very short and very crooked legs. Now it occurred to its owner—his name was Seth Wright—that if he could get a stock of sheep like this, they could be much more easily kept within bounds than the ordinary breed; and so, as soon as the young ram arrived at maturity, he killed his old ram and bred altogether from the deformed one. In a very few years

II.

III.

he was able to get together a very considerable flock of these deformed
sheep, or "ancons" as they were called, and they were spread all through
Massachusetts. Unfortunately, about this time the merino sheep were
introduced, and as they were a quiet race of sheep, and showed no tendency
to trespass or jump over fences, and as they carried much better wool than
the ancons, these latter were allowed to die out.

You see that these facts illustrate perfectly well what may be done if you
take care to breed from stocks that are similar to each other. After having
got a variation, if, by crossing it with the original stock, you multiply that
variation, and then make the variations breed together, then you may
almost certainly produce a race whose tendency to continue the variation
is exceedingly strong. This is what is called "selection"; and it is
exactly by the same process as that by which Seth Wright bred his ancon
sheep, that all our breeds of cattle, and dogs, and fowls are obtained.

Now the next question that lies before us is this : Does this selective
breeding occur in nature? Is there any principle of *natural* selection?
Because, if not, then all that I have been telling you goes for nothing in
accounting for the origin of species. And this brings me to the consideration of the Conditions of Existence, in which I think I shall be able to show
you that we have a cause competent to play the part of "selection" in
perpetuating varieties.

By Conditions of Existence I mean two things—there are conditions
which are furnished by the physical, the inorganic world, and there are
conditions furnished by the organic world. There is, in the first place,
Climate. Under this head I include only temperature and the varied
amount of moisture in particular places. In the next place there is what is
technically known as "Station," which means, given the climate, the
particular kind of place in which the plant or animal lives. For example,
the station of a fish is the water; the station of a marine fish is the sea, and
a marine fish may have a station higher or lower. So again with land
animals, and with plants, the differences in their stations are those of soil
and neighbourhood, and so forth. The third condition of existence is Food ;
by which I mean food in its broadest sense, the chemical materials
necessary to support the life of any organised being. Then come the
organic conditions; by which I mean the conditions which depend upon
the state of the rest of organic creation, upon the number and kind of the
living organisms by which a plant or animal is surrounded. You may
class these under two heads : there are organic beings which operate as
opponents, and there are organic beings which operate as helpers to any
given organism.

Now, how do the Conditions of Existence act the part of the "fancier"
by selecting certain varieties for perpetuation. I can give you the answer
in a few words ; but fully to understand their meaning is not an easy task.
They create the "Struggle for Life," and as this principle is such an
important one, I will write that also upon the board. The Struggle for
Life. Let me try to give you some idea how this Struggle for Life is
brought about, and how it acts as a selecting cause in the fixing of
variations. Imagine, if you can, a desert island, with a uniformly fertile
soil in which nothing whatever is growing ; and suppose that on this
island is dropped by a passing sea-bird a single seed, say of wheat, which
grows up and bears, perhaps, twenty other seeds, and that these in their
turn produce the same number, and so on. It is manifest that in a few
years they will have covered the entire island, and left no room for another

IIII

A. ①. ②. ③.

B. ①. ②.

"   "

plant. Now let us see what will happen in the following year. The time has arrived, as it will arrive for all organic things, when the birth-rate and the death-rate must be equal. As before, each plant will bear twenty seeds; but there will only be room for one, and so it is nineteen chances against one that any particular seed will survive. Now suppose that, as would be sure to happen, some of these seeds vary a little, a very little, from the others. Suppose that one of them is a little larger and stronger than the others, or that its integument is a little thinner and will allow it to germinate a little sooner. It is obvious that the seed which has the most favourable variation will have the best chance of life. Therefore, the seeds which survive to produce the crop of the following year, will be a little stronger or a little earlier than their parents; and this will go on year by year, the variation becoming fixed by heredity until some other cause intervenes. Now here we have the action of only one of the Conditions of Life. If we suppose that the other conditions are operating, the result will be the same. Imagine, for instance, that the sea-birds deposit more guano on one end of the island than the other; you will see that the crops at that end would gradually become of finer quality than those at the other. Or again, suppose that bees were introduced into the island. They would be direct helpers of the plants in carrying the fertilizing pollen from one to the other. So that if a plant varied one year in such a way as to be slightly better adapted to the wants of the bees than the rest, it would be better fertilized; and this variation, by giving the plant a better chance in the struggle for life, would be naturally selected for perpetuation. You now see, I hope, how the Conditions of Existence, by which all organic things are surrounded, give rise to that Struggle for Life in which they are engaged; and you will also see, and this brings me to my last point, that in this Struggle for Life we have that principle of Natural Selection for which we were enquiring, and that, given the existence of organic matter, we can now see how it may have been modified into an infinite number of widely different forms.

III.—Composition and Foreign Languages.

In writing in Shorthand short passages in a foreign language, such as may be expected to occur in office work, in note-taking, or in literary composition, it is not necessary to represent all the sounds exactly as they are pronounced. Such a practice would entail the learning of an entirely new form of the system, adapted to the language in question. It is sufficient as a rule to represent the language as if the words were sounded according to the natural English pronunciation, and trust to the writer's knowledge of the language to read them correctly. Thus in French *temps* might be written ⁄, and *ton* ⁄⁄; though in the adaptation of the system to French, in which ⁄ represents the nasal, the words would be represented ⁄⁄ (t o ng), and ⁄⁄ (t a͟w ng). It is on these principles that the following passages are written :—

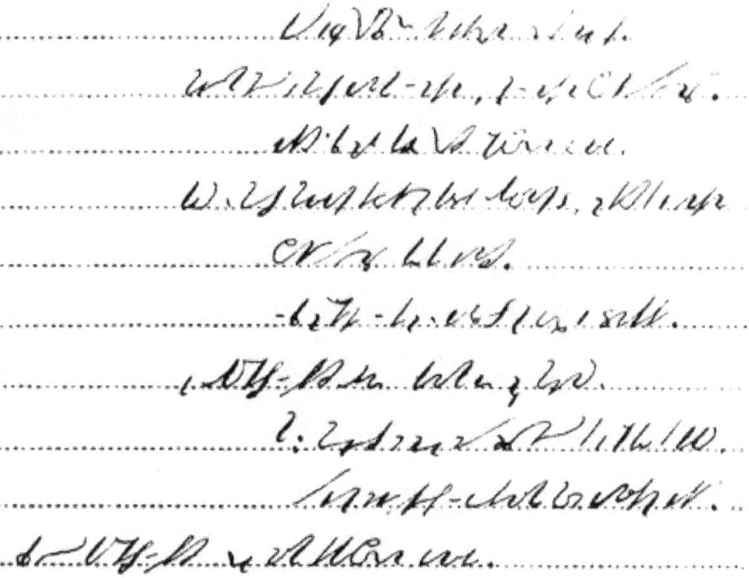

V.

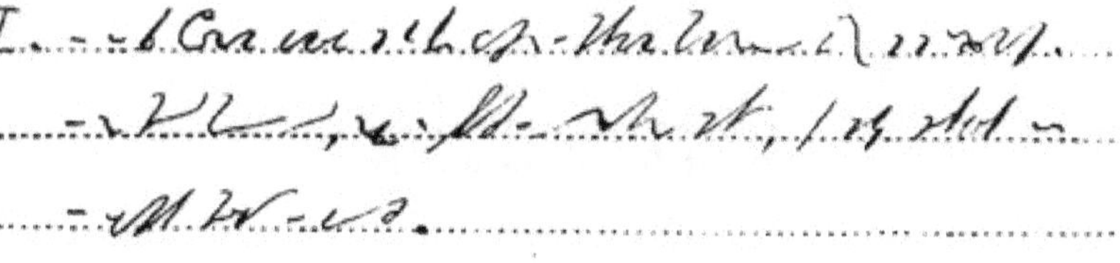

III

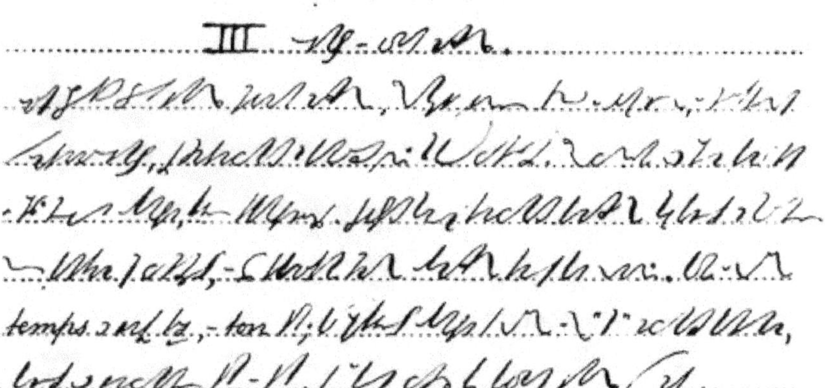

A.—LATIN.

HORACE.　Carm. i. 9.

Vides, ut alta stet nive candidum
Soracte, nec jam sustineant onus
　　Silvæ laborantes, geluque
　　　　Flumina constiterint acuto.

Dissolve frigus, ligna supor foco
Large reponens, atque benignius
　　Deprome quadrimum Sabina,
　　　　O Thaliarche, merum diota.

Permitte Divis cetera, qui simul
Stravere ventos æquore fervido
　　Deprœliantes, nec cupressi
　　　　Nec veteres agitantur orni.

Quid sit futurum cras, fuge quærere, et
Quem Fors dierum cunque dabit, lucro
　　Appone, nec dulces amores
　　　　Sperne puer neque tu choreas,

Donec virenti canities abest
Morosa.　Nunc et campus et areæ
　　Lenesque sub noctem susurri
　　　　Composita repetantur hora,

Nunc et latentis proditor intimo
Gratus puellæ risus ab angulo
　　Pignusque dereptum lacertis,
　　　　Aut digito male pertinaci.

B.—ENGLISH COMPOSITION.

TRANSLATION OF THE ABOVE.

See how in robe of deepest white
　Soracte stands, the woods below
　Are bending 'neath their load of snow,
Ice-bound the rivers stay their flight.

Pile high the logs, the fire shall shine,
　And drive the winter's cold away,
　While with more liberal hand to-day
We quaff deep draughts of Sabine wine

A.

B.

For in the Gods to trust is best.
Who soon shall still the winds again,
Now battling on the stormy main,
And give the tossing forests rest.

Seek not to know what is to be,
But, as the days on silent wing
Fly by, whate'er their hours may bring,
Count that a blessing falls to thee ;

Nor while thy youthful days remain,
Thy joyous locks are raven yet,
Young love's sweet dalliance forget,
Nor quit the dance with sour disdain ;

But to the Campus oft repair,
To try thy skill, or prove thy power,
Or at the evening trysting hour,
Hold whispered converse with the fair.

Then shalt thou hear the tell-tale cry,
When from the slyly shrinking maid,
In some far corner's welcome shade,
Thou snatch'st a ring in memory.

A. J. C.

C.—GREEK.

AESCH. CHOEPH. 576-591.

Πολλὰ μὲν γᾶ τρέφει δεινὰ δειμάτων ἄχη,
   πόντιαί τ᾽ ἀγκάλαι
   κνωδάλων ἀνταίων
βρύουσι· πλάθουσι καὶ πεδαίχμιοι
   λαμπάδες πεδάοροι·
πτανά τε καὶ πεδοβάμον᾽ ἀπ᾽ ἀνεμοέντων
   αἰγίδων φράσαι κότον.

ἀλλ᾽ ὑπέρτολμον ἀνδρὸς φρόνημα τίς λεγοι
   καὶ γυναικῶν φρεσὶν
   τλημόνων παντόλμους
ἔρωτας ἄταισι συννόμους βροτῶν ;
   ξυζύγους δ᾽ ὁμαυλίας
θηλυκρατὴς ἀπέρωτος ἔρως παρανικᾷ
   κνωδάλων τε καὶ βρυτῶν.

C. 576.591

D.—FRENCH.

Zola. La Débacle. V.
Charge of the French Cavalry at Sedan.

Alors, le colonel du premier régiment, levant en l'air son sabre, cria d'une voix de tonnerre :

—— Chargez !

Les trompettes sonnaient, la masse s'ébranla, d'abord au trot. Prosper se trouvait au premier rang, mais presque à l'extrémité de l'aile droite. Le grand danger est au centre, où le tir de l'ennemi s'acharne d'instinct. Lorsqu'on fut sur la crète du calvaire et que l'on commença à descendre de l'autre côté, vers la vaste plaine, il aperçut très nettement, à un millier de mètres, les carrés prussiens sur lesquels on les jetait. D'ailleurs, il trottait comme dans un rêve, il avait une légèreté, un flottement d'être endormi, un vide extraordinaire de cervelle, qui le laissait sans une idée. C'était la machine qui allait, sous une impulsion irrésistible. On répétait : "Sentez la botte ! Sentez la botte !" pour serrer les rangs le plus possible et leur donner une résistance de granit. Puis, à mesure que le trot s'accélérait, se changeait en galop enragé, les chasseurs d'Afrique poussaient, à la mode arabe, des cris sauvages, qui affolaient leurs montures. Bientôt, ce fut une course diabolique, un train d'enfer, ce furieux galop, ces hurlements féroces, que le crépitement des balles accompagnait d'un bruit de grêle, en tapant sur tout le métal, les gamelles, les bidons, le cuivre des uniformes et des harnais. Dans cette grêle, passait l'ouragan de vent et de foudre dont le sol tremblait, laissant au soleil une odeur de laine brûlée et de fauves en sueur.

E.—EVIDENCE.

Action for Breach of Promise.
Cross-examination of the Defendant.

[In the Reporting Style.]

Q.—How long have you known the plaintiff ?
A.—Since 1880, anno domini.
Q.—Since childhood, in fact?
A.—No, not since childhood.
Q.—Why, how old are you ?
A.—I think that is rather a rude question to ask.
Q.—How old are you ?
A.—I believe I was twenty-three years old last birthday.

D. _____ ____, _____ .5.

E. _____.

_____ . [_____].

_____   8 00 _____

_____   7-8 _____

Q.—Then in 1880 you would not have been more than thirteen years old?
A.—I suppose so.
Q.—Did you not walk out with the plaintiff for fourteen months?
A.—I don't know.
Q.—Well, did you walk out with her for fourteen days?
A.—I dunno.
Q.—You cannot swear whether it was fourteen days or fourteen months?
A.—I expect it was months.
Q.—Why did you walk out with her?
A.—I dunno.
Q.—Did you write her letters during that time?
A.—Yes.
Q.—Did you call her " darling " ?
A.—Yes, but I never proposed to her.
Q.—When you sent her " fondest love and best wishes for her future career," what did that refer to?
A.—I don't know.
Q.—It did not refer to her career as your wife?
A.—No.
Q.—Why did you ask her to bring you all the past correspondence?
A.—Because it was our custom to give back our letters.
The Court.—I suppose all correspondence was marked : " To be returned to the owner." *(Laughter.)*

THE END.

BEMROSE & SONS, LTD., PRINTERS, DERBY AND LONDON.—52049.